Principles of Animal Research for Graduate and Undergraduate Students

Principles of Animal Research for Graduate and Undergraduate Students

Mark A. Suckow
University of Minnesota, MN, USA

Kay L. Stewart
University of Notre Dame, IN, USA

AMSTERDAM • BOSTON • HEIDELBERG • LONDON
NEW YORK • OXFORD • PARIS • SAN DIEGO
SAN FRANCISCO • SINGAPORE • SYDNEY • TOKYO

Academic Press is an imprint of Elsevier

Academic Press is an imprint of Elsevier
125 London Wall, London EC2Y 5AS, United Kingdom
525 B Street, Suite 1800, San Diego, CA 92101-4495, United States
50 Hampshire Street, 5th Floor, Cambridge, MA 02139, United States
The Boulevard, Langford Lane, Kidlington, Oxford OX5 1GB, United Kingdom

Notices

Knowledge and best practice in this field are constantly changing. As new research and experience broaden our understanding, changes in research methods, professional practices, or medical treatment may become necessary.

Practitioners and researchers must always rely on their own experience and knowledge in evaluating and using any information, methods, compounds, or experiments described herein. In using such information or methods they should be mindful of their own safety and the safety of others, including parties for whom they have a professional responsibility.

To the fullest extent of the law, neither the Publisher nor the authors, contributors, or editors, assume any liability for any injury and/or damage to persons or property as a matter of products liability, negligence or otherwise, or from any use or operation of any methods, products, instructions, or ideas contained in the material herein.

Library of Congress Cataloging-in-Publication Data
A catalog record for this book is available from the Library of Congress

British Library Cataloguing-in-Publication Data
A catalogue record for this book is available from the British Library

ISBN: 978-0-12-802151-4

For information on all Academic Press publications
visit our website at https://www.elsevier.com/

Working together
to grow libraries in
developing countries

www.elsevier.com • www.bookaid.org

Acquisition Editor: Sara Tenney
Editorial Project Manager: Fenton Coulthurst
Production Project Manager: Edward Taylor
Designer: Greg Harris

Typeset by TNQ Books and Journals

Contents

Contributors

L.C. Anderson
Global Animal Welfare and Comparative Medicine Covance Laboratories, Madison, WI, United States

K. Bayne
AAALAC International, Frederick, MD, United States

M.J. Cramer
University of Notre Dame, Notre Dame, IN, United States

J.R. Crisler
Indiana University, Indianapolis, IN, United States

M.C. Dyson
The University of Michigan Medical School, Ann Arbor, MI, United States

D.L. Hickman
Indiana University, Indianapolis, IN, United States

J. Johnson
The University of the West Indies, Trinidad, West Indies

N.A. Johnston
Indiana University, Indianapolis, IN, United States

G.A. Lamberti
University of Notre Dame, Notre Dame, IN, United States

K.D. Prongay
Oregon National Primate Research Center, Beaverton, OR, United States

S. Putta
University of Southern California, Los Angeles, CA, United States

J. Robichaud
University of Notre Dame, Notre Dame, IN, United States

K.L. Rogers
Indiana University, Bloomington, IN, United States

R. Shepherd
Indiana University, Bloomington, IN, United States

K.L. Stewart
University of Notre Dame, Notre Dame, IN, United States

M.A. Suckow
Veterinary Population Medicine Department, University College of Veterinary Medicine, St. Paul, MN, United States

T.H. Vemulapalli
Purdue University, West Lafayette, IN, United States

Preface

Of the many interactions we have with animals, their use in research is particularly complex. Animals have been used to study basic physiological processes, as well as to expand our knowledge of important diseases, including cancer, infectious disease, diabetes, and many others. Indeed, it is widely accepted that most major advances in medicine were achieved through the use of animals as models. Still, ethical consideration must be given when undertaking research work that will involve the use of animals. The overall ethical cost of the animals used and their experience as research subjects is often weighed against the potential benefit to humans, as well as animals, in terms of improved health and veterinary care. Against this backdrop, it is essential that the use of animals in research be judicious and responsible, and that the handling and care of animals be responsible and compassionate. Toward this end, training of personnel, including students, who will work with research animals is key; and it is with that in mind that we undertook creation of this book. An important goal was to develop a text that would provide an overview of relevant topics for students, particularly undergraduate students.

Principles of Animal Research provides an overview of topics related to animal research featuring chapters on basic aspects of research animal care and use, including basic biology and behavior of common species, the ethics and regulatory oversight of animal research, the effect of variables on research outcomes, common technical procedures, and safety considerations for working with animals. In addition, chapters have been included to address field studies conducted using animals and thesis development for the undergraduate student. Our objective is not to present these topics in an exhaustive manner, but, instead, to offer students and others enough of an introduction to allow responsible undertaking of animal research; and it is our hope that this book will prove to be a useful basic resource for those seeking to learn and understand the responsibility that accompanies the privilege of working with animals in research.

Mark A. Suckow, DVM, DACLAM
University of Minnesota
and
Kay L. Stewart, RVT, CMAR
University of Notre Dame

Brief Historical Overview on the Use of Animals in Research

1

M.A. Suckow[1], K.L. Stewart[2]

University of Minnesota, Minneapolis, MN, United States[1]; University of Notre Dame, Notre Dame, IN, United States[2]

CHAPTER OUTLINE

1.1 INTRODUCTION

Animals play significant and varied roles in our lives, often to our benefit though sometimes to our disadvantage. For example, pet animals serve as loyal companions; some animals provide entertainment, such as a dog retrieving a tossed object; some perform work, such as herding, guarding, or retrieving; and some provide food. In contrast, our interactions with animals are sometimes viewed as negative, such as with verminous rodents or aggressive animals that inflict injury. Nonetheless, the animal kingdom has developed an overall close, often symbiotic, relationship with man.

One particular way in which the existence of man and animals has intersected is the use of animals as research and teaching subjects. The advancement of knowledge through the use of animals has benefited both human and animal health through scientific inquiry. It is widely understood that while differences do exist and there is no perfect model, extrapolation of knowledge learned from animal studies can be used to better understand the physiology, anatomy, and disease processes of man. This chapter will explore how the particular relationship between man and animals as research subjects has developed over time.

1.2 EARLY ATTITUDES TOWARD ANIMAL HEALTH AND DISEASE

Proper perspective of early attitudes requires some understanding of current thinking. The idea that animals represent sound models which allow researchers to gain insight into basic principles of biology and medicine and which can be applied to solving problems of mankind is widely accepted by the scientific community. Underlying this principle is recognition that basic biologic processes and disease processes are similar, though not necessarily identical, between animals and humans, thus supporting the thinking that animals can serve as research models. The term, **comparative medicine**, refers to the study of phenomena that are basic to diseases of all species. Similarly, the term **zoobiquity** is sometimes used to describe the idea that humans and animals get many of the same diseases and that communication and interaction between those specializing in human and animal health should logically yield synergy and benefit for both man and animals.

1.2.1 EARLY RELATIONSHIP OF MAN WITH ANIMALS

At approximately 20,000—15,000 BC, man changed from being strictly a vegetarian to a nomadic hunter. The relationship of man as the hunter and animals as prey slowly evolved to incorporate symbiosis with some animals. For example, gradually man would share the kill of dog packs and eventually began hunting in unison with dogs. This relationship led to domestication of dogs and later domestication and husbandry of sheep, goats, and cattle by 3000 BC. Over time, domestication and animal husbandry became an established part of civilization. That animals are susceptible to disease was noted early, as plagues often decimated domesticated animal populations, thus impacting humans who depended on livestock for food and labor (Wilkinson, 2005).

The early philosophers weighed in on the relationship of man to the animal kingdom with varied opinions and perspectives. Plato had a rather low opinion of the animal world and discouraged any sort of comparison between man and animals (Wilkinson, 2005). In contrast, Celsus and Galen both used animals to better understand anatomy and physiology as it might relate to man, with Galen using a variety of species, including mice, sheep, pigs, dogs, weasels, bears, and on one occasion an elephant.

1.2.2 ANIMAL DISEASE AS A WAY TO LEARN ABOUT HUMAN DISEASE

In the fifth century, AD, Vegetius recorded observations and thoughts with respect to the care of domestic animals, including the care of cattle and horses ill with infectious disease, such as the viral disease, rinderpest and the bacterial disease, strangles. This was driven in no small part by the economic importance of these species at that time. Sometimes referred to as the "Veterinary Hippocrates," Vegetius postulated the idea that much disease was the result of transmitted contagion; that is infectious agents. This was in contrast to the widely held belief that such illness in animals was the result of divine displeasure. From this idea arose methods to control infectious disease of animals via isolation and it was not long before comparisons were made to similar diseases in humans.

In the mid-13th century, Albertus Magnus wrote *De animalibus*, a treatise that discussed human and animal diseases together, and postulated ideas related to transmission of animal diseases; that is, they might be transmitted via bite; via contact with a diseased animal; or via air respired from a sick animal. These ideas led to further unification of theories underlying diseases of animals and man.

The application of the principles of isolation as an approach to control infectious disease in both man and animals continued and gained importance throughout Europe. In 1775, having successfully led an effort in France to control a rinderpest outbreak in cattle, Fèlix Vicq d' Azyr expressed the idea that just as there is a great deal to be learned about man through comparative anatomy, there is also a great deal to be learned from "comparative medicine." In this way, then, d' Azyr first articulated the idea that animals could be used to systematically study diseases with import to man. Further, d' Azyr believed that, given the unity of human and animal medicine, animals could be used to conduct important experiments that would be considered criminal if conducted in humans.

Over time, animals came to be seen as important tools in efforts to more clearly understand diseases affecting man. Early efforts focused on the study of infectious disease. For example, in the 1800s, Francois Magendie studied the transmission of rabies, first by inoculation of saliva from an affected human to a dog. Investigations by others into "wool sorter's disease," now referred to as anthrax, used rabbits, suckling pigs, guinea pigs, as well as dogs, to establish a transmissible, bacterial agent as the cause of the disease.

In 1901, John D. Rockefeller established the Rockefeller Institute for Medical Research in New York. The underlying idea was to link scientific research to clinical teaching within a medical school. Interestingly, the public was encouraged at that time to bring sick animals to the Institute for evaluation. In 1910, a Plymouth Rock hen was brought to the Institute and came to the attention of Dr. Peyton Rous, a physician. Subsequent work resulted in isolation of the first retrovirus and led to Rous being awarded the Nobel Prize in Medicine in 1966.

By the mid-20th century, the idea that animals can serve as important models to understand the basis of human disease and to develop treatments became an accepted part of scientific practice. Even today, concepts that are developed through the use of nonanimal technologies, such as molecular biology, are frequently evaluated in animals to establish validity in an organismic system. This is especially true for technologies that are hoped to be used as treatments for disease in humans, since animal testing is typically required before any evaluation in humans can begin.

1.3 DEVELOPMENT AND EMERGENCE OF NEW ANIMAL MODELS

As the use of animals as research subjects became a more common and standard part of the scientific process, the animals used became somewhat more standardized. For their small size, relative ease of maintenance, high fecundity, and relatively short lifespan mice and rats emerged as the species that were most commonly used. Systematic breeding of rodents led to inbred strains of animals with specific, defined characteristics that could be leveraged for some types of inquiry. For example, the Brattleboro rat was developed at a private research laboratory in West Brattleboro, Vermont when an astute animal care assistant noticed that drinking bottles attached to one cage among hundreds of Long-Evans rats were nearly always empty. Upon learning of this observation, the scientist in charge of the rats recognized the importance of the discovery bred animals from this particular line and developed them as an animal model of diabetes insipidus, an important human disease involving hypothalamic control of water intake (Valtin, 1982). Another such serendipitous example involving recognition of an animal model of an important human disease is the Lobund-Wistar rat. In 1973, Morris Pollard noted at the Lobund Institute of the University of Notre Dame that several Wistar rats which were part of a breeding colony maintained for germfree research had developed prostate cancer. Recognizing the possibility that this might represent a model system which could be used to study the same disease of man, Pollard developed a breeding colony of these animals that was characterized by frequent, spontaneous prostate cancer which metastasized to the lungs (Pollard and Suckow, 2005). Through this process of serendipity, keen observation, and selective breeding, many unique lines of rats and mice were developed as models of disease.

More recently, technology has sped the development of new animal models. In particular, molecular biology techniques have aided creation of new genetic models. Beginning largely in the 1980s, the transfer of specific genes to the genomes of mice, and more recently other organisms, has allowed production of "transgenic" models with modified characteristics resulting from the genetic manipulation (Pinkert, 2002). In some cases, transgenic models demonstrate clinical disease similar to a disease of humans, and in other cases they express other characteristics of scientific interest such as dramatically increased growth resulting from overexpression of growth hormone (Palmiter et al., 1982). Since the original experiments with transgenic mice, a wide variety of techniques have been developed for many species that allow addition, deletion, or other modifications of genetic elements within the genomes of animals. As a result, there exists now great possibility for the study of disease in animals where no such opportunity previously existed.

Because some important biological features of humans are not readily replicated in animal models, there has been great interest in development of "humanized" models; that is, animal models that have distinct human elements as part of their makeup. For example, human tissues and cells can be implanted into immunodeficient mice. In this way, implanted human hepatocytes have been used to study aspects of human liver function (Grompe and Strom, 2013). Further, immunodeficient mice that have been irradiated and then implanted with human blood forming cells can be infected with viruses and bacteria as a way to study the immunologic response to such pathogens (Leung et al., 2013). It is also possible to use molecular biology techniques to create genetically humanized mice which carry human DNA sequences in some or all of their cells (Scheer et al., 2013). In this way, it has been possible to create models for study of hereditary diseases of man, as well as models that allow investigations into drug efficacy, cancer, and infectious disease (Scheer et al., 2013).

1.4 IMPROVEMENTS IN APPROACHES TO ANIMAL CARE

The foundation of sound animal care is to provide the animals with an environment that consistently meets their biological needs and the needs of the research. This means that strategies should be used that provide a safe environment from which they cannot readily escape or are exposed to hazards, as well as meet the biological requirements for comfort in terms of feed, bedding, and environmental quality. Further, animal care should strive to reduce variables in the environment, as inconsistency can lead to spurious research results.

1.4.1 CAGES

Lacking specific standards, early cages, such as those for rodents, were constructed from wood sealed with varnish and covered with a wire top (Hessler, 1999). Over time, it was learned that such cages were prone to chewing by the mice housed therein and were not easily sanitized as the varnish became worn. Further, consumption of varnish by the animals represented an additional risk to the animals and possible experimental variable.

Over time, other materials replaced wood. For example, first galvanized and later stainless steel was used to construct cages for larger animals, such as dogs, cats, and nonhuman primates. These same materials were used to construct racks to hold cages of smaller animals, such as rodents. Due to ease of corrosion, stainless steel became more widely used and is still common today (Fig. 1.1).

Eventually, plastics, such as polycarbonate and polystyrene, replaced wood as the material of choice for construction of rodent cages (Fig. 1.2). Of these, polycarbonate has proved to be the most widely used, with many including durability and the ability to withstand autoclaving. Versus wood, rodents generally do not chew on plastic cages, and the cages are nonporous and easily sanitized.

The standard rodent cage was generally placed on a rack and covered with a wire bar lid. However, infectious disease of rodents was a common factor that complicated the health of the animals as well as their research value. Over time, it became recognized that infectious disease was a serious factor in the successful use of

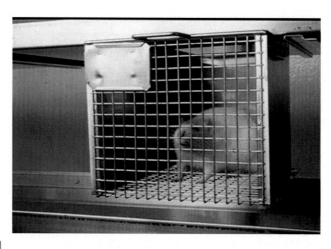

FIGURE 1.1

A single albino rat housed in a cage made of stainless steel wire. The cage requires the animal to stand on wire, a circumstance that may result in a condition known as ulcerative pododermatitis and which involves open sores on the bottom of the feet of heavy rats housed in this way for extended periods of time. For this reason, inclusion of a solid resting platform is encouraged when such cages are used.

FIGURE 1.2

Several albino rats housed in a polycarbonate cage that is equipped with bedding on the floor and a red plastic shelter for environmental enrichment. The shelter allows the animals opportunity for interaction with a novel device, as well as providing a shelter beneath which they can rest.

animals as research models. In the case of rodents, cages were developed in which a filter was placed over the top of the cage so that air could flow in and out, but particles that might carry infectious organisms were filtered out. Because of the somewhat restricted airflow through these filters, such cages have a tendency for moisture and ammonia to increase. More recently, individually ventilated caging (IVC) systems have been developed in which the cages having a filter top are connected to a rack that directly feeds filtered air into the cages, thus creating positive air pressure within the cage so that air from the outside (and, possibly particles carrying infectious organisms) cannot flow inward to the animals (Fig. 1.1). IVC cage systems are now widely employed and considered to be an important tool in the maintenance of rodents that are free from specified pathogens. In addition, the continuous airflow maintains the cages in such a way that moisture and ammonia levels are reduced compared to those equipped with only a static filter (Fig. 1.3).

1.4.2 FOOD AND WATER

Feed provided to animals is an important consideration in their care, as the nutritional quality is essential to animal health. Further, inconsistent nutritional content

FIGURE 1.3

Cages of mice on an individually ventilated caging rack. At the top of the rack is a high efficiency particulate air filter that is part of a system that allows delivery of air that is relatively free of particulate contaminants to the cages on the rack.

represents a variable that can affect experimental outcomes. For example, it has been demonstrated that commercial animal diets containing soy proteins may contain isoflavones, compounds which influence hormonal activity with consequent potential to artificially modify experimental results (Brown and Setchell, 2001). In the early 1970s, a program was established to standardize laboratory animal diets (Barnard et al., 2009). This program made it possible to closely control diets fed to research animals, thereby ensuring adequate nutrition and minimization of diet as a variable.

Similar to feed, the water provided to research animals has come to be recognized as an important factor in ensuring animal health and experimental integrity. The quality of water provided to animals is now recognized as important, with attention given to potability. In general, municipal water supplies are regarded as adequate for most circumstances, though some situations that require extra effort to reduce the possibility of infectious agents or contaminants being passed via the drinking water require purification strategies such as reverse osmosis. Though a variety of methods, including bowls and glass bottles with sipper tubes, are still used as a means to deliver water to the animals, automatic watering systems that provide a water source directly plumbed to the animals' cage or pen are now common and employed as a means to reduce the labor associated with filling a large number of bottles with water.

1.4.3 ENVIRONMENTAL PARAMETERS

Conditions for housing research animals did not at first receive great consideration. The main concern was to have sufficient space to allow production of, and housing for, enough animals to meet the needs of the research. Animals were housed in whatever space was available and, in the absence of established guidelines, felt to be adequate. As a result, the conditions under which animals were housed varied greatly and challenged the ability of researchers to achieve consist, reproducible results.

In 1950, the Animal Care Panel (ACP) was organized by a group of five veterinarians who had responsibility for the care of laboratory animals. One goal of the ACP was to see that standards were developed for the care of research animals. In the early 1960s, the ACP was awarded a contract from the National Institutes of Health to "determine and establish professional standards for laboratory animal care and facilities" (Mulder, 1999). The work of the ACP resulted in publication of the *Guide for the Care and Use of Laboratory Animal Facilities and Care* in 1963. Successive revisions have led to the eighth edition of this document, published in 2011 (National Research Council, 2011).

Presently known as the *Guide for the Care and Use of Laboratory Animals*, or more simply the *Guide*, the document defines a standard that is based largely on professional judgment. Within the *Guide*, principles and standards are provided for a variety of areas, including program oversight, training and occupational health of personnel, veterinary care, physical plant of animal facilities, and animal environment and housing. With respect to animal environment, standards are provided for relevant parameters, such as temperature, ventilation, humidity, illumination, noise, cage and social environment, and sanitation.

1.5 SUMMARY

As the relationship of man with animals has evolved, so has the recognition of the value presented to research by the use of animal models. In many ways, this relationship has been symbiotic, as knowledge gained from the use of animals in research has aided and improved not only the wellbeing of man, but also of animals. Drugs, devices, and methods used to enhance the health of man have in many cases also found use in veterinary medicine.

With the increasing recognition of the importance of controlling variables associated with the care of animals and an appreciation for the responsibility to meet basic biological needs of animals used in research, emphasis has been placed on establishing standards for care. With confidence, one can predict that as scientific inquiry expands our need to use animals as research subjects and increase our knowledge regarding the biological requirements of species used, our approach and perspective will no doubt evolve in tandem.

REFERENCES

Barnard, D.E., Lewis, S.M., Teter, B.B., Thigpen, J.E., 2009. Open- and closed-formula laboratory animal diets and their importance to research. J. Am. Assoc. Lab. Anim. Sci. 48, 709−713.

Brown, N.M., Setchell, K.D.R., 2001. Animal models impacted by phytoestrogens in commercial chow: implications for pathways influenced by hormones. Lab. Invest. 81, 735−747.

Grompe, M., Strom, S., 2013. Mice with human livers. Gastroenterol. 145, 1209−1214.

Hessler, J.R., 1999. The history of environmental improvements in laboratory animal science: caging systems, equipment, and facility design. In: McPherson, C.W., Mattingly, S.F. (Eds.), 50 Years of Laboratory Animal Science. American Association for Laboratory Animal Science, Memphis, pp. 92−120.

Leung, C., Chijioke, O., Gujer, C., Chatterjee, B., Antsiferova, O., Landtwing, V., McHugh, D., Raykova, A., Munz, C., 2013. Infectious diseases in humanized mice. Eur. J. Immunol. 43, 2246−2254.

Mulder, J.B., 1999. Creation and development of AALAS programs. In: McPherson, C.W., Mattingly, S.F. (Eds.), 50 Years of Laboratory Animal Science. American Association for Laboratory Animal Science, Memphis, pp. 12−15.

National Research Council, 2011. Guide for the Care and Use of Laboratory Animals, eighth ed. The National Academies Press, Washington, DC.

Palmiter, R.D., Brinster, R.L., Hammer, R.E., Trumbauer, M.E., Rosenfeld, M.G., Birnberg, N.C., Evans, R.M., 1982. Dramatic growth of mice that develop from eggs microinjected with metallothionein-growth hormone fusion genes. Nature 300, 611−615.

Pinkert, C.A., 2002. Introduction to transgenic animal technology. In: Pinkert, C.A. (Ed.), Transgenic Animal Technology. Academic Press, London, pp. 3–12.

Pollard, M., Suckow, M.A., 2005. Hormone-refractory prostate cancer in the Lobund-Wistar rat. Exp. Biol. Med. 230, 520–526.

Scheer, N., Snaith, M., Wolf, C.R., Seibler, J., 2013. Generation and utility of genetically humanized mouse models. Drug Discov. Today 18, 1200–1211.

Valtin, H., 1982. The discovery of the Brattleboro rat, recommended nomenclature, and the question of proper controls. Ann. N. Y Acad. Sci. 394, 1–9.

Wilkinson, L., 2005. Animals and Disease: An Introduction to the History of Comparative Medicine. Cambridge University Press, Cambridge.

Philosophical and Ethical Foundations

2

K.L. Rogers
Indiana University, Bloomington, IN, United States

CHAPTER OUTLINE

2.1 BASIC APPROACHES
2.1.1 TRADITIONAL VIEW

Our relationship with animals has changed significantly over the last 100 years or so. Traditional views of man's interaction with animals are derived from our position of dominance over animals as proposed through some Western religions. These views were further refined by philosophers, such as Thomas Aquinas and Immanuel Kant, who argued that we should not abuse animals, thus sparking the movement for protections against animal cruelty. More recent perspectives of animal ethics emerged such as contractarian morality, utilitarian morality, and the animal rights view which now substantially influence our societal view on the use of animals.

Several thousand years ago, our ancestors crossed the threshold toward domestication of animals as they learned how to use animals for work, food, warmth, clothing, travel, and companionship. Christian, Jewish, and Muslim believers saw nature as a reflection of divine creation with humans as the centerpiece. Humans were created in God's image and likeness to have power over all creatures of the earth. Humans were charged with serving as guardians or stewards of the animals since they viewed animals as having inherent value beyond their usefulness to humans. During the period up to the 19th century, taking an animal's life was widely acceptable as long as no suffering was involved in the animal's death, and the death was necessary for some human end.

2.1.2 ANTICRUELTY VIEW

The view that animals are there for our use as our needs dictate predominated until approximately the 19th century. During this time, ethical and political changes were occurring which made it no longer acceptable for the ruling classes to treat the lower classes in the way they treated their property. Humans were all thought to be equals with emphasis on their safety and happiness. Populations were making the shift from rural settings to urban settings with greater levels of wealth and more time to think about animal welfare. Thoughts shifted toward provision of happiness and safety for animals and toward limits on humans with respect to what they were allowed to do with, or to, animals in their care.

The first law for the protection of animals was passed in England in 1822 which said:

> *if any person or persons having the charge, care or custody of any horse, cow, ox, heifer, steer, sheep or other cattle, the property of any other person or persons, shall wantonly beat, abuse or ill-treat any such animal, such individuals shall be brought before a Justice of the Peace or other magistrate*

Ryder (1989)

Passage of the law was somewhat of a compromise with numerous exclusions. The species protected by the law included those used in entertainment "sports" at the time, which included cock fighting, dog fights, and bull or bearbaiting (in which dogs attacked a chained bull or bear) (Fig. 2.1). With the popularity of such sports, politicians were reluctant to make laws that made them illegal. In addition, animal species were also thought to belong to a moral hierarchy with animals being rated on how useful they were, how closely one worked with the animal, how affectionate the animal was, and how harmful the animal could be. At this time, horses and cattle were at the top of the hierarchy, followed by dogs and cats which were kept as pets. Animals at the lower end of the hierarchy included rats and mice that were considered to be pests. Animals at the top of the scale were to be treated with greater regard. Notable in this law was the fact that it only protected animals against acts done by people who were not the owners. The owners were presumed to want to protect their animals for use in food, fiber, transportation and farm labor.

2.1.3 THE DEVELOPMENT OF CONTRACTARIANISM

During the 19th and 20th centuries, various theories on the appropriate treatment and interaction with animals began to emerge. Initially, animals were

Bear-Baiting.

FIGURE 2.1

Bearbaiting—a form of blood sport once popular in Europe, and still practiced in some parts of the world, in which a tethered bear would fight a number of dogs. In this engraving from late 18th century England, things are out of control because the bear has gotten loose. In 1835 bearbaiting was banned in England because it involved "wanton cruelty."

Engraving reproduced from Brand, J., 1841. Observations on Popular Antiquities, London. Found in Sandoe, P., Christiansen, S.B., 2008. Ethics of Animal Use. Blackwell Publishing, Oxford, p. 4.

seen as resources for humans to use, and thus humans were allowed to treat animals in ways they would not treat humans. The thought was that humans were justified to treat animals the way they wanted because animals mattered less than human beings. During the early 19th century, philosophers such as Jeremy Bentham (1748–1832) began to question the moral status of animals. He noted that some humans had lower intellectual capacity and linguistic ability (such as babies or the mentally impaired) that was considered below the ability of some animals. Intelligence and linguistic ability were therefore not the only reasons humans should be considered morally superior to animals. Bentham proposed that maybe the capacity to suffer which confers moral status, thus implying humans and animals to be of similar moral status.

According to the contractarian perspective, one thinks or acts morally because it is in one's self-interest. Thus, *when one is obliged to show consideration for other people this is really for one's own sake. In general, by respecting the rules of morality one contributes to the maintenance of a society that is essential to one's own welfare. Moral rules are conventions which best serve the self-interest of all members of the society. Contractarian morality applies only to individuals who can 'contract in' to the moral community, so it is important to define who these members are: 'On the contract view of morality, morality is a sort of agreement among rational, independent, self-interested persons, persons who have something to gain from entering into such an agreement…'*

Sandoe and Christiansen (2008)

According to the contractarian view, there is a difference between the moral status of humans and animals. Humans act with respect and cooperate with other humans; and those humans act similarly in return because of a "contract" agreed upon by the participants. However, animals will not react in similar terms to humans because they cannot enter into such a contract. In the view of the contractarian John Rawls, as contractors (participants), we should ignore the characteristics that make us different as humans to eliminate the possibility of excluding certain groups of humans from the contract. Rawls believed that infants and mentally disadvantaged humans, for example, cannot enter into rational contracts and therefore would deny that we owe direct duties to those humans. This view leads to Rawls prejudice of speciesism in reference to animals since they also cannot enter into contracts. Contractarians, therefore, assign greater weight to human interests than animal interests simply because they are human interests. Humans are only expected to treat animals well enough to be used for the purpose intended by humans. In contrast, humans have direct duties to other humans whose interests are morally relevant. However, while animals have interests, their interests are not of direct moral concern.

2.1.4 UTILITARIANISM

The philosophical views of Bentham and Singer reflect those of utilitarianism. For the utilitarian, every individual's interests count morally and deserve equal consideration and are affected by the action of another. For this view, an interest or preference is defined in terms of "the capacity for suffering and/or enjoyment or happiness" (Regan and Singer, 1989). Individuals, therefore, prefer acts that will enhance their enjoyment or reduce their suffering. Further, all sentient beings, including humans and nonhumans, have interests that count morally. Thus, any actions imposed on a sentient being, including animals, would need to take into account the interests of all parties involved. In terms of our relationship with animals, this would mean we should take into account animals' interests and look toward improvement of their welfare. Specifically, Peter Singer said:

As long as a sentient being is conscious, it has an interest in experiencing as much pleasure and as little pain as possible. Sentience suffices to place a being within the sphere of equal consideration of interests; but it does not mean that the being

*has a personal interest in continuing to live. For a non-self-conscious being, death
is the cessation of experiences, in much the same way that birth is the beginning of
experiences. Death cannot be contrary to a preference for continued life, any more
than birth could be in accordance with a preference for commencing life…Given
that an animal belongs to a species incapable of self-consciousness, it follows that
it is not wrong to rear and kill it for food, provided that it lives a pleasant life and,
after being killed, will be replaced by another animal which will lead a similarly
pleasant life and would not have existed if the first animal had not been killed. This
means that vegetarianism is not obligatory for those who can obtain meat from animals that they know to have been reared in this manner*

Singer (1993)

The utilitarian, then, is one who judges actions to be right or wrong based on the
consequences or states of affairs they bring about. The state of affairs that is best is
one which produces a maximum of pleasure over pain. In this view, a "sentient
being" is one that is capable of awareness or consciously experiencing pleasure or
suffering. Plants, for example, do not possess the ability to feel pleasure or pain,
are not conscious, and therefore are not sentient beings. Utilitarianism considers
things we know and value, such as avoiding pain and seeking pleasure. According
to conventional morality, most people would choose an action that results in the
greatest good for the majority of individuals and this is in alignment with utilitarianism. Moral judgments for utilitarians require universalizability so that decisions
or actions should be taken that bring about the best balance between all preferences
and all frustrations for everyone affected. Utilitarians would, therefore, consider the
interests of every being affected by an action and give them the same weight as other
beings with an interest in the situation. Therefore, with regard to animal research,
utilitarians such as Singer believe that the pleasure and pains of animals used in
research deserve consideration equal to those of humans in calculating the overall
benefits of animal experimentation. Singer believes that the researchers should be
required to demonstrate that the benefits of their research would outweigh the
suffering of the animals involved.

2.1.5 ANIMAL RIGHTS VIEW

As the developed world grew in wealth and population, the use of animals for
food expanded. As food production became more commercialized, and efficiencies maximized, any focus on animal welfare became less significant and
fell into the background. Animals were placed in smaller and more efficient
cage systems to maximize food production efficiency. Increased efficiency
resulted in reduced production cost that often came at the expense of freedom
of movement for the animals in smaller, more restrictive cages.

In 17th century England and France, animals were used to understand both animal
and human anatomy and physiology (Fig. 2.2). Animal experimentation also played a
significant role in the development of vaccines and treatments for such things as anthrax,

FIGURE 2.2

Illustration by Andreas Vesalius. Pig tied to dissection board "for the administration of vivisections."

From Saunders, O'Malley, 1950. Taken from Loew, F.M., Cohen, B.J., 2002. Laboratory Animal Medicine: Historical Perspectives. In: Laboratory Animal Medicine, second ed., Elsevier, p. 2. RightsLink permission obtained 12/16/14.

smallpox, rabies, yellow fever, typhus, and polio. As the use of animals in research increased, people started taking notice of the use and treatment of animals in research. This awareness prompted legislation for the protection of these animals.

Over time, attitudes changed, and people began to value the companionship provided by animals. Dogs and cats became popular pets that were not generally killed unless they developed serious illness. The protection and preservation of wild animals grew in interest. Veterinary practice evolved from treatment of agricultural animals to companion animal practice and the development of veterinary subspecialties. These trends led to growing opposition to the use of animals for anything other than companionship. The most extreme of these movements regarding animal use, the **animal rights** movement, opposes most common forms of animal use, including research. These groups believe that animals have more than simple moral status; that is, that animals have rights. Animal rights groups began championing the effort to protect animals from being used for food and instead adopting vegetarian or vegan diets. Other groups focus on specific goals such as putting an end to commercial whaling, hunting of sea mammals, or banning fur production. Some of the more extreme groups have performed acts of terrorism on institutions or the individual researchers performing the work in the name of protecting the animals and ending use of animals in research. In the animal rights view, it is always unacceptable to treat a sentient being merely as a means to obtain a goal. They believe that all humans have intrinsic worth or dignity and should be treated as an end and never as a means, and they have extended this principle of dignity and preservation of rights to animals.

2.2 ANIMAL WELFARE VERSUS ANIMAL RIGHTS
2.2.1 ANIMAL WELFARE

As described earlier, many believe that sentient animals have interests that are significant by themselves. The utilitarian believes that we should enable animals to fulfill their behavioral needs and protect them from suffering. In contrast, the animal rights viewpoint holds that we are to maintain the animal's interests or rights. What are an animal's needs and preferences? To answer this, a more scientific approach can be taken with consideration of animal physiology, veterinary medicine, and animal behavior. Animal welfare science looks to objectively and scientifically identify the needs of animals and determine if they are being met.

An animal's welfare can be measured by the use of three parameters: health, physiology, and behavior. We often ascertain the health of an animal in terms of the amount of illness the animal has experienced. An animal that has more signs of illness or an increase in the severity of sickness experiences greater levels of suffering or discomfort. Animal welfare as defined by Donald Broom is as follows:

The welfare of an individual is its state as regards its attempts to cope with its environment. Coping can sometimes be achieved with little effort and expenditure of resources, in which case the individual's welfare is satisfactory. Or it may fail to cope at all, in which case its welfare is obviously poor. Or, if the individual does cope with the conditions it encounters, this may be easy, with little expenditure of resources, or may be difficult taking much time and energy, in which case welfare is deemed to be poor

Broom and Fraser (2007)

Physiological measurements can also provide insight into an animal's welfare. Measures such as heart rate, respiration, and blood pressure can indicate an animal's response to a situation or stimulus. A rise in heart rate, respiratory rate or blood pressure may indicate an increase in activity level of the animal. However, while an animal is at rest, an increase in those parameters may indicate that the animal is having trouble adapting to the situation and is stressed. Animals can also respond to stress with increases in hormones such as adrenalin, noradrenalin, and corticosteroids. Since adrenalin and noradrenalin are both broken down by the body over a period of hours, they are useful measures for acute stress. Corticosteroids, because they are released more slowly from the adrenal cortex, are suitable measures of animal stress over longer periods of time. There are frequently a number of factors affecting these physiologic measures such as an increase in activity leading to both an elevation in heart rate and increases in noradrenalin that would not be associated with stress in healthy animals.

In contrast, behavior can be evaluated through observation of the animal's movements and actions in real time or analysis of video recordings. Animals should be observed in a natural or seminatural environment which allows the observer to become familiar with the typical behavioral pattern of the animals. In the natural environment, one can observe behaviors associated with foraging, play, social and maternal interactions, and other types of behavior as a way to establish what is normal for that animal. The animal's normal behavior can then be compared to behaviors displayed in different environments or under different circumstances to assess the animal's welfare. Over time, animals have been selectively bred for purposes related to farming, companionship, and research and had the ability to adapt gradually to these situations. Animal welfare problems arise when the animals are kept in conditions to which they no longer can adapt. Animals placed in these new environments may develop frustration that is manifested as abnormal behavior, such as stereotypies (e.g., bar biting in sows or circling in dogs) or redirected behavior (e.g., feather pecking in chickens). This abnormal behavior is an indication that the animal is no longer able to adapt and is experiencing impaired or poor welfare.

Various tests have been developed to assess more objectively an animal's response to "normal" or controlled conditions versus altered conditions. One of these tests is the **choice test** where the animal is placed in a situation with access to two different conditions and asked to choose which one it prefers. An example might be a rat that is placed between a small cage and a larger cage to see which one it "prefers." The other test is an operant conditioning test to determine the amount of effort the animal makes to gain access to a particular resource or to perform a particular behavior. For example, an animal might be taught to press a lever for a food reward. Gradually, the number of presses needed to obtain the food reward increases. If the reward is good enough, the animal may continue to press the lever to obtain the reward; however, if the task becomes too hard (pressing the lever too many times), the animal may stop pressing the lever to get the food rewards. However, in both of these situations, the choice the animal makes can also be influenced by experiences earlier in life, e.g., the animal may choose the smaller cage because it has the type of bedding the animal used early in life.

Because no one parameter can be relied upon to provide unequivocal evidence about an animal's welfare, it is important to evaluate several different behavioral parameters to determine an animal's welfare. For example, a single dog placed in a cage might be seen to have a decreased appetite and found to be frequently walking in circles within the cage. This same dog is taken to a cage twice the size and placed with another dog. The original dog does not circle, its appetite improves and the dogs are occasionally seen playing with each other. In this situation, it would be fair to say that the animal's welfare was improved when it switched from single housing to pair housing because its appetite returned to normal, it stopped a stereotypical behavior (circling) and it was seen to be playing with its cage mate. Physiological measures such as heart rate and respiratory rate, if taken, would most likely also indicate that when paired and at rest, the dog's heart rate and respiratory rate would be decreased indicating improved animal welfare.

2.2.2 ANIMAL RIGHTS

The term "animal rights" can be used in three different contexts. A common interpretation leads the majority of the public to believe that animals have rights. This same public also believes that humans have the right to kill and eat animals, thereby denying animals their right to life. In this sense, they believe that animals have the right to some, rather limited, moral consideration and protection from abuse and maltreatment. The term "animal rights" can also be used in a political sense to lend some political credibility to a movement whereby people can campaign for increased consideration for animals and their enhanced moral status. "Animal rights" are also sometimes invoked in an effort to eliminate the use of animals in research, industry, agriculture, and entertainment. When used philosophically, a "right" is a claim that can only be overridden by another "rights claim." Carl Cohen states that "a right, including a moral right, is a claim or potential claim that one party may exercise against another" (Cohen, 1986). One argument for animal rights states that animals have the right to not suffer and therefore should not be used in animal research where they may suffer. In addition, some have strengthened the argument by stating that we cannot use animals merely as a means to our own end and that animals have their own inherent value.

Tom Regan, a philosopher who supports the animal rights movement's aim to eliminate the use of animals by humans completely (the abolitionist approach), is one of the more extreme proponents of the animal rights view. Cohen describes the animal rights view about animal experimentation this way:

> *Animal experimentation, they say, along with the eating of animals and every other disrespectful use of animals, is to be condemned not conditionally but absolutely, not because it does more harm than good but because it is intrinsically and absolutely wrong*

Cohen and Regan (2001)

Regan builds his argument in this way. He believes that individuals who have interests have inherent value. An individual with inherent value is one who has inherent worth along with equal moral status; individuals are not merely of instrumental value. Those individuals who possess inherent value are owed the direct duty of respectful treatment or to be treated as *somebodies* rather than *somethings*. This duty of respect is owed to all persons regardless of sexual orientation, age, race, gender, or intellectual abilities. In this nonspeciesist view (the view that being a member of a particular species does not confer unique rights), Regan also recognizes that direct duties are owed to animals also. What is important for Regan in the animal rights view is how a person or animal is treated, and if treated without respect, as a thing or without morally significant value, the treatment is morally wrong. As Regan states,

> *Using my terminology, to murder the innocent for this reason is to treat those who possess inherent value as if they are of instrumental value only—as if their moral status was the same as a pencil or skillet, a pair of roller skates or a Walkman…recognition of the inherent value individuals share and the direct duty of respect owed to those who have it…*

Cohen and Regan (2001)

Participants in the animal rights movement believe that humans and animals have rights. As defined by Cohen,

> *A right is a valid claim, or potential claim, that may be made by a moral agent, under principles that govern both the claimant and the target of the claim. Every genuine right has some possessor and must have some target and some content*

Cohen and Regan (2001)

A right or claim, then, represents treatment that one is justified to demand from another; treatment from another that is owed or is a duty to the individual.

People, therefore, have the right to be treated with respect, not to be harmed or interfered with. People have the right to life, bodily integrity, and freedom. However, according to some philosophical views, not all humans are persons, including the newly fertilized ovum or a permanently comatose human. Regan tries to include these individuals by raising the concept of "subject-of-a-life." Regan believes each "subject-of-a-life" brings to its life consciousness with awareness and feelings, beliefs, and desires. People as individuals live an experiential life with feelings, desires, and beliefs pulled together into one psychological individual with distinct thoughts, actions, and experiences unique to the individual. In Regan's words,

All those who satisfy this criterion—that is, all those who, as subjects-of-a-life, have an experiential welfare—possess inherent value—thus are owed the direct duty to be treated with respect, thus have a right to such treatment

Cohen and Regan (2001)

Regan states that animals, such as mammals and birds, are sentient beings and share cognitive, attitudinal, emotional, and volitional capacities with humans. In support of this, he notes that animals resemble humans in their behavior, physiology, anatomy, and psychology and have some evolutionary relationship to humans. Humans and animals are all subjects-of-a-life with inherent value and rights. What is not clear from Regan's views is where the line is drawn between animals with rights and those without. For example, he would agree that an insect does not have rights, but it is unclear at what level nonhuman animals become subjects-of-a-life and gain rights.

Philosophical background aside, many animal rights views have developed over the last 40 years. Animal rights groups may rally around a specific situation, calling for the end of the use of fur, catching of whales, using nonhuman primates in research, and commercial production of eggs or chicken through confined housing systems. Other groups such as People for the Ethical Treatment of Animals (PETA), Stop Animal Exploitation Now (SAEN), Physicians Committee for Responsible Medicine (PCRM), and even the Humane Society of the United States (HSUS) are specifically opposed to the use of animals in research. PETA's mission statement reinforces their animal rights focus.

> " People for the Ethical Treatment of Animals (PETA) is the largest animal rights organization in the world, with more than 5 million members and supporters. PETA works to stop all animal cruelty and focuses its attention on the four areas in which the largest numbers of animals suffer the most intensely, for the longest periods of time: on factory farms, in the clothing trade, in laboratories, and in the entertainment industry."

Mission statement of the People for the Ethical Treatment of Animals (PETA).
From PETA, 2016. About PETA/Our Mission Statement. http://www.peta.org/about-peta/.

The HSUS appears to take a less rigid approach to animal research.

> "We're working to decrease and eventually end the use of animals in testing, research and teaching by promoting the development of innovative and effective alternative methods.
> For the past few decades, the HSUS has conducted its campaigns promoting alternatives and the reduction of animal use in a way that avoids harassment of individual scientists. The organization has declined to participate in campaigns focused against individuals and feels strongly that violence, illegal action, or the harassment and intimidation of animal researchers or any other category of people who use animals is unacceptable and cannot be condoned as society debates and moves away from the use of animals in."

Statement taken from the "Animals in Laboratories" section of Humane Society of the United States (HSUS) Website.

Humane Society of the United States, 2016. Animals in Research. http://www.humanesociety.org/about/ departments/animals_research.html.

Some animal rights groups have allegedly paid operatives to obtain pictures and associated information from research organizations which, after editing, make it erroneously appear that animals are being abused in research (Fig. 2.3).

There have also been some animal rights activists on the edges of the movement who have become violent in response to what they perceive to be abusive and immoral practices. Animal rights groups such as Animal Liberation Front (ALF) and Stop Huntingdon Animal Cruelty (SHAC) have been linked to aggressive tactics. In 2003, terrorists claimed responsibility for the detonation of explosive devices at Chiron Life Sciences Center and Shaklee Inc. because of their business links to Huntingdon Life Sciences. After the second bombing, an anonymous communiqué was issued,

> *We gave all of the customers the chance, the choice, to withdraw their business from HLS (Huntingdon Life Sciences). Now you will all reap what you have sown. All customers and their families are considered legitimate targets… You never know when your house, your car even, might go boom… Or maybe it will be a shot in the dark… We will now be doubling the size of every device we make. Today it is 10 pounds, tomorrow 20… until your buildings are nothing more than rubble. It is time for this war to truly have two sides. No more will all the killing be done by the oppressors, now the oppressed will strike back*

Lewis (2004)

Luckily there were no injuries in the explosions at Chiron and Shaklee; however, this signaled that animal rights activists were willing to abandon nonviolent approaches. Instead, some were willing to adopt more confrontational and aggressive tactics designed to threaten and intimidate legitimate companies into abandoning entire projects or contracts.

Animal Testing 101

Right now, millions of mice, rats, rabbits, primates, cats, dogs, and other animals are locked inside cold, barren cages in laboratories across the country. They languish in pain, ache with loneliness, and long to roam free and use their minds.

Instead, all they can do is sit and wait in fear of the next terrifying and painful procedure that will be performed on them. The stress, sterility and boredom causes some animals to develop neurotic behaviors such incessantly spinning in circles, rocking back and forth and even pulling out their own hair and biting their own skin. They shake and cower in fear whenever someone walks past their cages and their blood pressure spikes drastically. After enduring lives of pain, loneliness and terror, almost all of them will be killed.

More than 100 million animals every year suffer and die in cruel chemical, drug, food, and cosmetics tests as well as in biology lessons, medical training exercises, and curiosity-driven medical experiments at universities. Exact numbers aren't available because mice, rats, birds, and cold-blooded animals—who make up more than 95 percent of animals used in experiments—are not covered by even the minimal protections of the Animal Welfare Act and therefore go uncounted. To test cosmetics, household cleaners, and other consumer products, hundreds of thousands of animals are poisoned, blinded, and killed every year by cruel corporations. Mice and rats are forced to inhale toxic fumes, dogs are force-fed pesticides, and rabbits have corrosive chemicals rubbed onto their skin and eyes. Many of these tests are not even required by law, and they often produce inaccurate or misleading results. Even if a product harms animals, it can still be marketed to consumers. Cruel and deadly toxicity tests are also conducted as part of massive regulatory testing programs that are often funded by U.S. taxpayers' money. The Environmental Protection Agency, the Food and Drug Administration, the National Toxicology Program, and the Department of Agriculture are just a few of the government agencies that subject animals to painful and crude tests.

The federal government and many health charities waste precious dollars from taxpayers and generous donors on cruel and misleading animal experiments at universities and private laboratories instead of spending them on promising clinical, in vitro and epidemiological studies that are actually relevant to humans.

Millions of animals also suffer and die for classroom biology experiments and dissection, even though modern alternatives have repeatedly been shown to teach students better, save teachers time, and save schools money.

Each of us can help save animals from suffering and death in experiments by demanding that our alma maters stop experimenting on animals, by buying cruelty-free products, by donating only to charities that don't experiment on animals, by requesting alternatives to animal dissection and by demanding the immediate implementation of humane, effective non-animal tests by government agencies and corporations.

PETA's extensive worldwide campaign to expose and end the use of animals in experiments is wide-ranging. In addition to our well-known groundbreaking undercover work and colorful advocacy campaigns to educate the public, with the help of our members and supporters we push government agencies to stop funding and conducting experiments on animals; encourage pharmaceutical, chemical, and consumer products companies to replace tests on animals with more effective non-animal methods; help students and teachers end dissection in the classroom; fund humane non-animal research; publish scientific papers; and urge health charities not to invest in dead-end tests on animals. Our staff is working on these issues, and this multifaceted approach yields scores of victories for animals locked in laboratories every year.

FIGURE 2.3

An article about animals used in experimentation on the People for the Ethical Treatment of Animals Website, http://www.peta.org/issues/animals-used-for-experimentation/animal-testing-101/.

In 1992, the Animal Enterprise Protection Act (AEPA) is a law created to protect industries against the increasing number of violent attacks by animal rights extremists, and resulted due to perceived inadequate protection from state and federal laws protecting the "vital services" produced at these companies (Civil Liberties Defense Center, 2012). After additional animal rights activity in 2003, proponents of a new law emerged stating that the current law had not provided sufficient deterrence from animal rights threats and targeting of those affiliated with animal enterprises. The https://www.govtrack.us/congress/bills/109/s3880 (S. 3880-109th Congress: Animal Enterprise Terrorism Act, www.Gov.Track.us 2006, October 16, 2016) was passed by Congress in 2006 to prohibit any person from engaging in certain conduct "for the purpose of damaging or interfering with the operations of an animal enterprise." It covers any act that either "damages or causes the loss of any real or personal property" or "places a person in reasonable fear" of injury. The law also contains a clause that indicates it should not be construed to "prohibit any expressive conduct (including peaceful picketing or other peaceful demonstration) protected from legal prohibition by the First Amendment to the Constitution." The AETA gives the US Department of Justice greater authority to respond to animal rights threats by broadening the definition of "animal enterprise" to include academic and commercial enterprises that use or sell animals or animal products (Wikipedia, 2016). The law also increases penalties to take into consideration the amount of economic damage caused and allows affected animal enterprises to seek restitution.

2.3 OPPOSITION TO AND BENEFITS OF ANIMAL EXPERIMENTATION

2.3.1 OPPOSITION TO ANIMAL EXPERIMENTATION

Significant opposition to animal research arises from the erroneous perception that animal experimentation is not necessary to advancing human and animal health. The combined results of many projects, that might individually seem to yield only minor advancements, often lead to striking improvements in understanding causes, prevention, and treatment of disease. In the pharmaceutical industry, for example, a number of animals are used in experiments to discover and develop drugs that are slightly better, need to be taken less frequently, and have other desirable traits than currently existing drugs. These improved drugs may be beneficial because of their slightly enhanced efficiency or toxicity profile, but, strictly speaking, they may not be necessary to prevent, alleviate, or cure human disease. Part of the debate over the use of animals in research is regarding whether it is morally acceptable to use animals in developing treatments for conditions such as baldness, and the testing of cosmetics and other related luxury products.

There exist real dilemmas when it comes to testing of these "luxury products." Industries producing such personal enhancement products have an obligation to provide products that are not toxic on application and that do not cause allergies or cancer. However, these products are not necessary for society, and politicians feel the pressure of groups or individuals opposed to the use of animals in research to end such testing along with the pressures from companies wishing to produce such products. To address this in the European Union, there has been a substantial investment in the development of methods of safety testing that do not involve live animals, although some animal testing remains necessary. Another area where controversy exists in the use of animal testing is in research related to so-called lifestyle diseases, those caused by drinking, smoking, and overeating. Some of the induced diseases created by people partaking in these activities may be serious and widespread, but animal rights activists would argue that animals should not suffer because people cannot control themselves. In reality, very few lifestyle diseases are totally related to lifestyle. They may be the result of social factors, childhood conditions, or the combined result of genetic and environmental factors over which the individual has no control.

Use of animals in research is also sometimes opposed because animal models do not always identically mimic humans. As models, animals may provide additional insights into pathophysiology or disease, but they can also lead research astray. Opponents of animal research argue that each species has subtle but significant differences that cannot be predicted or fully understood to extrapolate to humans. For example, in a study looking at compounds that reduced ischemic stroke in rodents, none of the compounds were efficacious in human trials (Barnard and Kaufman, 1997). The lack of efficacy was potentially due to the difference between natural strokes that develop over time in humans versus the experimentally induced strokes produced in the rodents over a period of weeks. Opponents also claim that studies on monkeys delayed the development of a vaccine for poliomyelitis. Initial experiments conducted on primates revealed that the virus was attracted to neural tissue. Later, scientists found that the virus could be propagated in gastrointestinal tissue and was only neurotropic because the virus was administered to the monkeys through the nose. Subsequently, cell cultures from monkeys rather than humans were used for vaccine production exposing people to harmful monkey viruses (Barnard and Kaufman, 1997).

Opponents of the use of animals in research also oppose use of animals to test the safety of drugs or other compounds. Within the pharmaceutical industry, it was noted that of 19 chemicals known to cause cancer in humans when ingested, only seven caused cancer in mice and rats using standards set by the National Cancer Institute (Barnard and Kaufman, 1997). For example, an antidepressant, nomifensine, had minimal toxicity in rats, rabbits, dogs, and monkeys yet caused liver toxicity and anemia in humans. In these and other cases, it has been shown that some compounds have serious adverse reactions in humans that were not predicted by animal testing resulting in conditions in the treated humans that could lead to hospitalization, disability, or death.

Abolitionists, calling for an end to animal research, state that researchers have better methods available such as human epidemiological studies, clinical intervention trials, clinical observation aided by laboratory testing, autopsy studies, human tissue and cell cultures, endoscopic examination and biopsy, and imaging methods. Investigation of atherosclerotic heart disease in humans gained significant knowledge from the Framingham Heart Study, which began in 1948. This epidemiological study revealed risk factors for heart disease, including high cholesterol, smoking, and high blood pressure. Various factors were manipulated in controlled human clinical trials revealing that every 1% drop in serum cholesterol levels led to at least a 2% decrease in risk for heart disease (Barnard and Kaufman, 1997). Studies in heart disease patients further suggested that consuming a low-fat vegetarian diet, getting regular exercise, smoking cessation, and managing stress can reverse atherosclerotic blockages (Barnard and Kaufman, 1997).

2.3.2 BENEFITS OF ANIMAL EXPERIMENTATION

Scientists in the United States use approximately 30 million animals in research, of which around 29 million are rats, mice, birds, or fish (Speaking of Research, 2014). (Figs. 2.4, 2.5, and 2.6). Animal research is but one of several complementary approaches playing a crucial role in the development of modern medical treatments. One of the most notable discoveries in infectious disease research was made by the

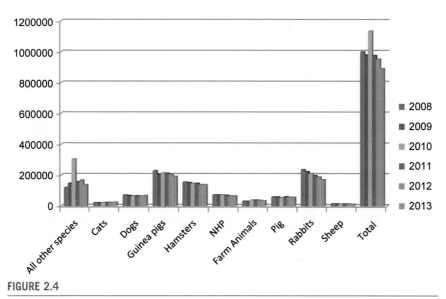

FIGURE 2.4

USDA covered species of animals used in research from 2008 to 2013 with each year depicted by a different color.

Information from http://www.aphis.usda.gov/wps/portal/aphis/ourfocus/animalwelfare/!ut/p/a1/04_
Sj9CPykssyOxPLMnMzOvMAfGjzOK9_D2MDJOMjDzd3V2dDDz93HwCzL29jAyCzYAKIvEo8DYITr-
zu6OHibmPgYGBiYWRgaeLk4eLuaWvgYGnGXH6DXAARwNC-sP1o_
AqAfkArACfE8EK8LihIDcONMIgOxMAwhVB1g!!/?1dmy&urile=wcm%3apath%3a%2FAPHIS_Content_Library
%2FSA_Our_Focus%2FSA_Animal_Welfare%2FSA_Obtain_Research_Facility_Annual_Report%2F.

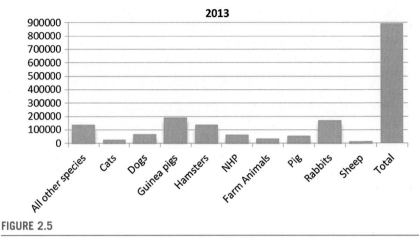

FIGURE 2.5

USDA covered species of animals used in research for 2013 from USDA annual reported numbers. http://www.aphis.usda.gov/animal_welfare/downloads/7023/Animals%20Used%20In%20Research%202013.pdf.

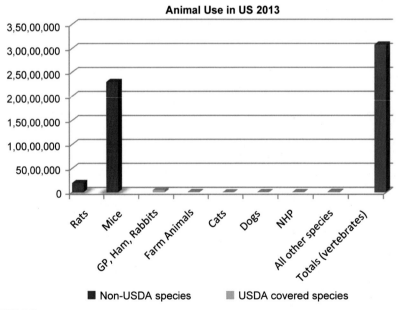

FIGURE 2.6

Total numbers of animals used for 2013 from published USDA annual numbers and estimates of mice and rats used. http://www.aphis.usda.gov/animal_welfare/downloads/7023/Animals%20Used%20In%20Research%202013.pdf.

French chemist, Louis Pasteur. By using animals to study infectious disease, Pasteur discovered the infectious agent causing avian cholera, *Vibrio cholerae*. He isolated microbes from chickens suffering from cholera and grew a possible causative microbe in culture. When he administered contents of the culture to healthy chickens, they also developed disease. He was even able to show that administration of cultures that had lost their ability to infect chickens was able to provide immunity against challenge with infectious cultures of organisms (Botting and Morrison, 1997). Pasteur's work influenced the practices of Joseph Lister, a British surgeon, who used carbolic acid to sterilize surgical instruments, sutures, and wound dressings. In an address to the Royal Commission, he stated that animal experiments had been essential to his own work on asepsis and that restricting research using animals would prevent discoveries that would benefit mankind (Botting and Morrison, 1997).

Ongoing research using animals has led to the development of a number of vaccines and drugs beneficial to human medicine. Scientists have determined the causative agent and developed vaccines for dozens of infectious diseases including anthrax, smallpox, diphtheria, tetanus, rabies, whooping cough, tuberculosis, poliomyelitis, measles, mumps, and rubella. Sir James Black, a British Nobel prize winner, used animals to develop beta blockers (Rollin, 2007). He discovered that hormones stimulate beta receptors in the heart resulting in an increased heartbeat and demand for oxygen. Black figured out a way to block the beta-receptor sites which prevented the hormones from attaching to them, resulting in a reduction in oxygen demand and thus reducing the pain of angina. When beta blockers were used, it was also observed that they produced a small but persistent decrease in blood pressure. Beta blockers were subsequently used to treat hypertension. With additional research, newer beta blockers were discovered in studies with rats to have fewer side effects (Coalition for Medical Progress, 2007). Breast cancer remains the most common cancer in women behind skin cancer. New drugs are being discovered to increase the possibility for prolonged survival. Aromatase inhibitors are drugs that block production of the female hormone, estrogen. One of these drugs, Herceptin (trastuzumab), was tested against the gold standard of breast cancer therapy, tamoxifen, by Professor Angela Brodie of the University of Maryland School of Medicine. Estrogen acts as a growth stimulator in some breast cancers and aromatase inhibitors, such as Herceptin, can be used to interfere with tumor growth in estrogen receptor—positive cancers (Coalition for Medical Progress, 2007).

Animal research has also been critically important to the treatment of heart disease and spinal cord injuries. A surgeon, John Heysham Gibbon, contributed significantly to the treatment of heart disease though development of the heart—lung bypass machine which was tried first in the cat. He continued to refine the machine in dogs until finally performing the first open heart operation on a human patient in 1953 (Coalition for Medical Progress, 2007). Prior to the heart—lung bypass machine, the average life expectancy for a patient with untreated heart valve disease was about 2 years. Using the heart—lung bypass machine in dogs allowed surgeons to operate on the heart while maintaining life through the bypass machine. This also allowed them to experiment with the heart valves of animals (pigs, sheep, calves, and goats) transplanted into dogs in the early 1970s to eventually be used as heart valves in humans. Researchers processed the animal valves to prevent rejection in the recipient and make the valves biologically inert. The first transplant of a pig heart valve into a human patient took place in 1975 and of a cow heart valve in 1981 (Coalition for Medical Progress, 2007). Current studies now research the use of artificial heart valves or biomaterials designed to serve as a scaffold for ingrowth of new cardiac cells resulting in the replacement of the biomaterial by functional heart tissue serving to patch heart defects or replace heart valves. Both mechanical (artificial) valves and valves taken from animals are used today.

People involved in traffic car accidents or climbing accidents can develop devastating spinal cord injuries where the links between the cord and the brain are severed. This trauma to the spinal cord can result in loss of sensation due to destruction of nerve fibers conveying sensory input to the brain, and loss of control of movement, breathing, bladder, bowels, sexual function, blood pressure, and body temperature. Using animals such as mice and rats, scientists can mimic human spinal cord damage. Nasal stem cells were discovered to regenerate themselves and were transplanted into paralyzed rats at the site of spinal cord injury and found to form a bridge across the injured spinal cord allowing the animals to walk and climb again (Coalition for Medical Progress, 2007).

2.4 PROFESSIONAL ETHICS RELATING TO BIOMEDICAL RESEARCH

The environment concerning the use of animals for research and safety testing has changed rather significantly, especially over the past 50 years, with increased levels of concern for animals by scientists and the public. Sentiment regarding animal research has been shaped as society has changed. As people increased their relationship with animals as companions, they started to consider the amount of good that would result from taking into account both human and animal interests and weighing them against the suffering, both human and animal, caused by the process as Peter Singer had done. Singer also introduced the concept of speciesism which states that we cannot ignore the interests of animals just because they are not human. Lastly, Tom Regan introduced the idea of "animal rights" and believed that all humans and most animals have inherent rights. He advocates for the abolition of animal use by humans altogether.

From the past philosophical and ethical thoughts on animal research, the public has developed their views and they remain inconsistent, although some recent polls indicate a decrease in support for animal research. These changes have coincided with society's shift from a high dependence on independently owned small farms to more urban lifestyles where the majority of people are shielded from the realities of raising agricultural animals for food. With increased urbanization has come the development of relationships with animals as pets or companions. This affection for keeping animals as pets has resulted in greater sympathy for animals and increased reluctance to see them experience pain and suffering in experiments.

Scientist and researchers' views have also evolved. Initially, researchers were taught that animals do not experience pain the way humans do and that they should learn to avoid emotional connections with the animals. Additional knowledge about animal physiology was gained in the 1990s when it was revealed that the pain sensing apparatus is the same throughout the vertebrate kingdom and that we should generally consider that if something causes pain to a human, it probably also causes pain to an animal. Increased regulation and legislation have led to more thorough oversight and control of the process for use of animals in experimentation.

Scientists' concern for animals first became widely apparent around the 1950s with the publication of *The Principles of Humane Experimental Technique* by a British zoologist, William M.S. Russell, and a microbiologist, Rex L. Burch. In this publication, Russell and Burch put forward the concept of "the three Rs" (**3 Rs**). The 3 Rs stated that research is best performed by taking into account the **R**eplacement of animals (e.g., by in vitro or test tube methods), **R**eduction of the numbers of animals used (e.g., by statistical techniques or other methods), and **R**efinement of the experiment (e.g., by using techniques with less potential to produce pain, suffering, or distress). In the 1960s, governments and humane organizations started to fund studies on alternative methods of animal research. With the increasing body of knowledge regarding alternative methods, organizations have developed to provide information to the research community on types of alternative procedures available. For example, the Center for Alternatives to Animal Testing (CAAT) at Johns Hopkins University was formed in 1981 in the United States to support the creation, development, validation, and use of alternatives to animal research, product safety testing, and education (Wikipedia, 2014). The Interagency Coordinating Committee on the Validation of Alternative Methods (ICCVAM) is a permanent committee of the National Institutes of Environmental Health and Safety (NIEHS) recognized in 2000 under the National Toxicology Program Interagency Center for the Evaluation of Alternative Toxicological Methods (NICEATM). ICCVAM is composed of representatives from 15 US Federal regulatory and research agencies that require, use, generate, or disseminate toxicological and safety testing information (National Toxicology Program, 2016). The European Society for Alternatives to Animal Testing (EUSAAT) encourages dissemination of information on alternative methods to animal testing, promotion of research to develop or validate 3 Rs' methods to replace, reduce, and refine animal tests, and the use of nonanimal tests in the area of education and continuing education (European Society for Alternatives to Animal Research, 2014). In the United Kingdom, the government established the National Center for the Replacement, Refinement and Reduction of Animals in Research (NC3Rs) in May 2004 to promote and develop high-quality research that takes the 3Rs into account (Festing and Wilkinson, 2007).

Implementation of the 3 Rs' concept has resulted in various methods to refine, reduce, or replace the use of animals in research and has resulted in some significant changes in the industry. The LD50 test was developed in 1927 for the biological standardization of dangerous drugs and was incorporated into the routine toxicological protocol for testing of chemicals (Zbinden and Flury-Roversi, 1981). This test determines the toxicity of compounds by using methods to establish the amount of a test agent required to kill 50% of animal test subjects. For scientific, economic, and ethical reasons, the LD50 test was reassessed and found to be inadequate because the precision of the procedure was dependent on the number of animals used. Even when large numbers of animals were used, there was considerable inherent variation in the test results due to many factors (animal species and strain, age, sex, diet, fasting versus provision of food, temperature, caging, season, experimental procedures, etc.). The test has been replaced by other methods that use fewer animals, and which have precision and reproducibility that are sufficient for most purposes of acute toxicity testing. Improved methods include physiological, hematological, biochemical, pathological, and histopathological techniques and simplified procedures using smaller numbers of animals, and make it possible to increase the amount of information gained from a study that does not require 50% of the animals to die from test compound toxicity. For the LD80 test used for vaccine testing, animals are vaccinated against the disease and then exposed to the disease agent. Vaccines are claimed to be effective if at least 80% of the vaccinated group remains healthy and 80% of the nonvaccinated control group dies (Mukerjee, 1997). It has since been determined that animals can be tested by checking their level of antibodies against the agent, and this only involves a small blood sample.

The field of bioinformatics has also contributed to the reduction in the use of animals for experimentation. For example, by reviewing decades of industry data on pesticides, it was determined that if mice and rats prove sensitive to a chemical, it does not have to be tested on dogs. Further, mathematical models capture the intricate interactions between RNA viral capsids, providing insights into viral evolution, infection, and assembly to assist with antiviral drug design or bionanotechnology. There are also efforts to combine nonanimal tests with computer programs that can predict skin allergy, acute systemic toxicity, and endocrine activity, among other things. Others are designing miniature, functioning 3-D liver cultures that will allow scientists to study how liver cells metabolize chemicals and drugs (Conlee, 2013).

In efforts to reduce or replace the use of animals in research, animals lower on the phylogenetic scale have been used along with the development of numerous in vitro techniques. As an in vitro replacement test for determining the corrosivity of chemicals, in 1999, Corrositex® was recommended as a stand-alone assay for evaluating acids, bases, and acid derivatives for the US Department of Transportation. In 2000, Corrositex® achieved formal acceptance by US regulators, and in 2006, the method achieved international acceptance (AltTox, 2010). EpiSkin, EpiDerm™, and the rat skin Transcutaneous Electrical Resistance (TER) tests were validated by European Union Reference Laboratory for Alternatives to Animal Testing (ECVAM) and endorsed in 1998 as validated methods to replace animal tests (AltTox, 2010). These tests were subsequently reviewed by ICCVAM, which recommended their use in a tiered testing strategy, followed in 2002 by European Union acceptance and international acceptance in 2004 (AltTox, 2010). The zebrafish (*Danio rerio*) has replaced the use of some transgenic rodents for advancing the understanding of genetic mechanisms and developmental processes, improving aquaculture, and for pharmaceutical discovery. The advantages of using zebrafish include their high fecundity, transparency of their embryos, rapid organogenesis, and availability of extensive genetic resources (Lee, 2014). Cryopreservation of sperm, oocytes, and embryos has allowed a reduction in the number of transgenic mice being bred to maintain genetically modified animals. Computers have also contributed to the reduction in animal use especially in the pharmaceutical industry where they are used to aid in drug design. Computers can help to determine the quantitative structure—activity relationships between physiochemical properties of the drug and its biological activity, which is used to predict potential for such things as carcinogenicity or mutagenicity.

Another source of change in the approach to the use of animals has been the result of changes in legislation. A federal law, the Animal Welfare Act (AWA), was introduced in 1966 in response to increased awareness of animal research and the idea that animals were being stolen from private owners for use in research. Later, the AWA was amended in 1985 following accusations of mistreatment of macaques at the Institute for Behavioral Research in Silver Spring, Maryland (Mukerjee, 1997). Though investigations by federal agencies later found little substance to the accusations, subsequent modifications of the AWA and the associated regulations introduced the requirement for an Institutional Animal Care and Use Committee (IACUC) as an internal review board for proposals to use animals in research, just as the institutional review boards review proposals to perform research in humans. This amendment also introduced a requirement to develop a plan for the exercise of dogs and psychological enrichment of nonhuman primates. The IACUC reviews proposed use of animals in research through submission of a standardized form which requests the Principle Investigator to describe the procedures that will be performed in animals, why he/she is conducting the study, the benefits to human or animal health, justification for the species of animals used and the number of animals, and the appropriate use of anesthetics and analgesics to reduce pain and distress. An animal use proposal involving pain or distress must include a search for alternatives to animal use that reduce, replace, or refine procedures. The investigator also must assure that the personnel performing the work on the animals have received appropriate training to perform the techniques listed in the protocol.

Lastly, the development and continuing evolution of laboratory animal medicine as a separate discipline within veterinary medicine has contributed to advancements in the research use of animals. Specifically, within veterinary medicine, individuals can partake in specialized training programs in laboratory animal medicine to become board certified by the American College of Laboratory Animal Medicine (ACLAM) (American College of Laboratory Animal Medicine, 2016). Such veterinarians have brought significant changes to the field by studying and discussing the specific needs and behaviors of laboratory animals to determine the best housing and husbandry conditions for the animals. Within the field, technicians can study and become certified through the American Association for Laboratory Animal Science (AALAS) at the technologist, technician, or assistant technician level. Those specialized in the operation of animal facilities can receive training through the Institute of Laboratory Animal Management (ILAM) to provide operational oversight of animal research facilities. They also use their administrative and technical skills to maintain and improve their animal care and use programs (American Association for Laboratory Animal Science, 2016). These training and certification opportunities have contributed to a more professional and ethical approach to use of animals in research.

In conclusion, our views and use of animals have evolved over time. Animals have been used by humans for food, fiber, work, and clothing as long as they have been treated in accord with their welfare and contractarianism. Utilitarianism, on the other hand, brings into consideration the interests of all parties involved. The animal rights view considers that animals have rights, including the right to life and should not be used in research pursuits even if it might be beneficial to human medical advancement, and in some cases that animals should not be used by humans at all. Medical history has provided evidence that animal research has been useful in the discovery of drugs, vaccines, improved surgical practices, and basic understanding of many diseases. As a whole, a range of viewpoints has allowed us to progress to a point where we now have laws, regulations, and guidelines that lead us to a very thoughtful consideration with respect to the use of animals in research. The IACUCs have been put in place to ensure that such consideration is made. As a whole, society will likely continue to evaluate our use of animals in research, and look for ways to reduce, replace, and refine such use.

REFERENCES

AltTox.org Non-Animal Methods for Toxicity Testing, 2010. Toxicity Endpoints & Tests: Skin Irritation/corrosion. http://alttox.org/mapp/toxicity-endpoints-tests/skin-irritationcorrosion/.

American Association for Laboratory Animal Science, 2016. General Information on Education/ILAM. https://www.aalas.org/education/ilam#.VJINxSvF8hw.

American College of Laboratory Animal Medicine (ACLAM), 2016. Training Programs. https://www.aclam.org/education-and-training/training-programs.

Barnard, N.D., Kaufman, S.R., 1997. Animal research is wasteful and misleading. Sci. Amer. 276, 80–82.

Botting, J.H., Morrison, A.R., 1997. Animal research is vital to medicine. Sci. Amer. 276, 83–85.

Broom, D.M., Fraser, A.F., 2007. Domestic Animal Behavior and Welfare, fourth ed. CABI, Cambridge, MA.

Center for Constitutional Rights, 2007. The Animal Enterprise Terrorism Act (AETA). http://ccrjustice.org/learn-more/faqs/factsheet%3A-animal-enterprise-terrorism-act-(aeta).

Civil Liberties Defense Center, 2012. The Redundancy of AETA. http://cldc.org/2012/01/09/aeta-the-redundancy-of-aeta/.

Coalition for Medical Progress, 2007. Medical Advances and Animal Research. The Contribution of Animal Science to the Medical Revolution: Some Case Histories. http://www.understandinganimalresearch.org.uk/files/7214/1041/0599/medical-advances-and.pdf.

Cohen, C., 1986. The case for the use of animals in biomedical research. New Engl. J. Med. 315, 865–870.

Cohen, C., Regan, T., 2001. The Animal Rights Debate. Rowman & Littlefield Publishers, Inc, Oxford.

Conlee, K., 2013. Technology Should Replace Testing on Animals. http://www.livescience.com/38326-new-approach-to-animal-testing.html.

European Society for Alternatives to Animal Research, 2014. EUSAAT: The European 3Rs Society. http://www.eusaat-congress.eu/index.php/eusaat.

Festing, S., Wilkinson, R., 2007. The ethics of animal research. Talking point on the use of animals in scientific research. EMBO Rep. 8, 526—530.

Humane Society of the United States, 2016. Animals in Research. http://www.humanesociety.org/about/departments/animals_research.html.

Lee, O., Green, J.M., Tyler, C.R., 2014. Transgenic fish systems and their application in ecotoxicology. Crit. Rev. Toxicol. 14, 1—18.

Lewis, J.E., 2004. Testimony before the Senate Judiciary Committee. http://www.fbi.gov/news/testimony/animal-rights-extremism-and-ecoterrorism.

Mukerjee, M., 1997. Trends in animal research. Sci. Amer. 276, 86—93.

National Toxicology Program, 2016. About ICCVAM. http://ntp.niehs.nih.gov/pubhealth/evalatm/iccvam/index.htmlhttp://ntp.niehs.nih.gov/pubhealth/evalatm/iccvam/index.html.

PETA, 2016. About PETA/Our Mission Statement. http://www.peta.org/about-peta/.

Regan, T., Singer, P., 1989. Animal Rights and Human Obligations. Prentice Hall, Englewood Cliffs.

Rollin, B.F., 2007. Animal research: a moral science. Talking point on the use of animals in scientific research. EMBO Rep. 8, 521—525.

Ryder, R.D., 1989. Animal Revolution. Basil Blackwell, Oxford.

Sandøe, P., Christiansen, S.B., 2008. Ethics of Animal Use. Blackwell Publishing, Oxford.

Singer, P., 1993. Practical Ethics, second ed. Cambridge University Press, Cambridge.

Speaking of Research, 2014. U.S. Statistics. www.speakingofresearch.com/facts/statistics/.

The Free Encyclopedia, 2016. Animal Enterprise Terrorism Act. http:Wikipedia//en.wikipedia.org/wiki/Animal_Enterprise_Terrorism_Act#cite_note-; law.cornell.edu-1.

Wikipedia, the Free Encyclopedia, 2014. Center for Alternatives to Animal Testing. http://en.wikipedia.org/wiki/Center_for_Alternatives_to_Animal_Testing.

Zbinden, G., Flury-Roversi, M., 1981. Significance of the LD50-test for the toxicological evaluation of chemical substances. Arch. Toxicol. 4 7, 77—99.

FURTHER READING

Cornell University Law School. Legal Information Institute, Open Access to Law Since 1992: 18 U.S. Code x 43-Force, violence, and threats involving animal enterprises. http://www.law.cornell.edu/uscode/text/18/43.

Doke, S.K., Dhawale, S.C., 2013. Alternatives to animal testing: a review. Saudi Pharm. J. 23, 223—229.

Gannon, F., 2007. Animal rights, human wrongs? Introduction to the talking point on the use of animals in scientific research. EMBO Rep. 8, 519—520.

Kraus, A.L., Renquist, D., 2000. Bioethics and the Use of Laboratory Animals. Gregory C. Benoit, Publishing, Dubuque.

Regan, T., 1983. The Case for Animal Rights. University of California Press, Berkeley.

Rowan, A.N., 1997. Forum: the benefits and ethics of animal research. Sci. Amer. 276, 79—93.

Regulations and Guidelines

3

K. Bayne[1], L.C. Anderson[2]

AAALAC International, Frederick, MD, United States[1]; Global Animal Welfare and Comparative Medicine Covance Laboratories, Madison, WI, United States[2]

CHAPTER OUTLINE

3.1 INTRODUCTION

In the United States, oversight of animal care and use for research, testing, and teaching is achieved by numerous laws, regulations, policies, and guidelines from two principal government organizations—the US Department of Agriculture (USDA) and the US Public Health Service (PHS). Other guidance may be derived from scientific panels [e.g., convened by the Institute for Laboratory Animal Research (ILAR), a component of the National Academies of Science] that are endorsed by the federal government as required standards or developed by professional associations that may be implemented by third-party audit systems [e.g., the Association for Assessment and Accreditation of Laboratory Animal Care International (AAALAC International)].

3.2 US ANIMAL WELFARE ACT

Since the 1966 Act, Congress has vested the USDA with both promulgation and enforcement authority over the use of animals in research. Research facilities, intermediate handlers, and common carriers are required to register with the USDA, while animal dealers and exhibitors must be licensed. Additional control over the quality of care afforded to animals used in research is achieved by the USDA requiring research facilities and government agencies to purchase animals only from USDA licensed sources, unless the source is exempted from obtaining a license. Through various methods of enforcement, the impact of the USDA's oversight responsibility is strong and has been used over the years to improve animal welfare at dealers, exhibits, and research facilities.

3.2.1 THE EVOLUTION OF THE ANIMAL WELFARE ACT

Animals used for research were first protected by federal legislation in 1966, with the passage of the Laboratory Animal Welfare Act (PL-89−544). The law was intended to address public concerns that pets were being stolen and sold to research facilities (NRC, 2009b). This concern was fueled by the story of Pepper, a Dalmatian who had gone missing from the yard of her home in 1965 and was ultimately euthanized as part of a research study, as reported in the November 29, 1965 issue of *Sports Illustrated* ("The Lost Pets that Stray to Labs"). This story was followed by an article published in *Life Magazine* on February 4, 1966, entitled "Concentration Camps for Dogs," which described the deplorable housing conditions at certain dog dealer facilities.

The Laboratory Animal Welfare Act required licensing of dealers (individuals or corporations) that bought or sold dogs or cats for research, as well as registration of research facilities that used dogs or cats. It also mandated minimum animal care standards for dogs and cats before and after they were used for research. Because the impetus for the law was to prevent unlawful trafficking of privately owned dogs and to ensure humane conditions for dogs at dealer facilities, the standards did not apply while the animals were being used for an experimental purpose. Further, a research facility only had to register if it received federal government funding and the dogs or cats purchased had crossed state lines. The law also authorized the USDA to develop and enforce regulations. The USDA subsequently established standards for nonhuman primates, rabbits, guinea pigs, and hamsters, in addition to those mandated for dogs and cats. Research facilities that used dogs or cats were required to observe the USDA-specified standards for all of these species. However, facilities that did not use dogs or cats were not required to comply with the standards for the other species.

The Laboratory Animal Welfare Act was amended in 1970 and renamed the Animal Welfare Act (AWA) (PL-91-579). The scope of protection was broadened to include animals used for teaching, exhibitions, and the wholesale pet industry. "Animals" included dogs, cats, nonhuman primates, rabbits, guinea pigs, and hamsters and, with certain exceptions, any other warm-blooded animals designated by the US Secretary of Agriculture. The animals no longer had to cross state lines to be covered by the law. Institutions (except primary and secondary schools) that used these species in research, tests, or experiments were required to register as a research facility. For the first time, zoos were required to be licensed. Specifically exempted were horses, which are not used in research, and agricultural animals used in food and fiber research, retail pet stores, state and county fairs, rodeos, purebred cat and dog shows, and agricultural exhibitions. The definition of "dealer" was revised to include any person who bought or sold any dog or other animal designated by the USDA for use in research, teaching, exhibition, or as a pet at the wholesale level.

The 1970 amendments also expanded the minimal animal care standards. These standards were applied to animals during the course of research, not only before and after experimental use. The Act did not allow the Secretary of Agriculture to establish rules, regulations, or orders with regard to the design or performance of the research. However, it required that every research facility submit an annual report that provided the number of regulated species it used and assurance that it met professionally acceptable standards for the care, treatment, and use of animals, including the appropriate use of anesthetic, analgesic, and tranquilizing drugs.

In 1976, the AWA was amended again (PL-94-279) to include regulation of common carriers and intermediate handlers and to establish transportation standards for animals. Carriers were required to be licensed. Standards were established for shipping conditions (e.g., feed, water, rest, ventilation, temperature, and handling) and for the containers in which animals were shipped. The amendments also prohibited interstate promotion or shipment of animals for animal fighting ventures.

In 1981, Alex Pacheco, who would subsequently become a cofounder of People for the Ethical Treatment of Animals, volunteered in a laboratory at the Institute for Biological Research in Silver Spring, Maryland, and filmed alleged violations of the AWA. As a result, the county police seized 17 monkeys and transferred custody of the animals to the National Institutes of Health (NIH). The incident, known as the Silver Spring Monkeys case, prompted congressional hearings (NIH, 1991). Ultimately, the Food Security Act of 1985 (PL-99-198) was passed. This bill included provisions to amend the AWA, referred to as "The Improved Standards for Laboratory Animal Act." These amendments were based on the following congressional findings:

1. The use of animals is instrumental in certain research and education for advancing knowledge of cures and treatment for diseases and injuries which afflict both humans and animals;
2. Methods of testing that do not use animals are being and continue to be developed which are faster, less expensive, and more accurate than traditional animal experiments for some purposes and further opportunities exist for the development of these methods of testing;
3. Measures which eliminate or minimize the unnecessary duplication of experiments on animals can result in more productive use of federal funds; and
4. Measures which help meet the public concern for laboratory animal care and treatment are important in assuring that research will continue to progress.

Until 1985, the AWA requirements and resulting USDA regulations were essentially limited to animal care, housing, and transportation standards. These new amendments included specific requirements for research facilities that were related to the experimental use of animals including consideration of alternatives to painful or distressing procedures, ensuring exercise of dogs, promoting the psychological well-being of nonhuman primates, and the establishment of an animal welfare information service within the National Agricultural Library (NAL). The law clearly states, however, that nothing in the Act should be construed as authorizing the Secretary of Agriculture to promulgate rules, regulations, or orders with regard to the design or performance of research protocols. It also mandates that the USDA may not interrupt the conduct of research during inspections. Importantly, and covered in detail in §2142(b), the Act mandated the establishment of Institutional Animal Care and Use Committees to oversee the use of animals in research, testing, and teaching at the local level.

The Pet Theft Act of 1990 (PL-101-624) was the fourth amendment to the AWA. It was incorporated in the 1990 Food, Agriculture, Conservation, and Trade Act, an omnibus farm bill and was referred to as the "Protection of Pets" legislation. The purpose of the amendment was to address reports of ongoing cases of pet theft, which resulted from gaps in the original law. This amendment established a 5-day holding period for dogs and cats held at pounds and shelters (both private and public) before releasing them to dealers. The holding period was designed to allow pet owners and prospective owners the opportunity to claim or adopt animals before they might be sold or used for research. In addition, the time period allowed appropriate documentation to be gathered regarding the source of the animal that would demonstrate the dog was obtained legally. The 1990 Act also allowed the USDA to seek injunctions against any licensed facility found dealing in stolen animals.

For many years a point of debate regarding the AWA centered on its exclusion of rats and mice. The 1970 amendment to the AWA stated that an animal was defined as "…any live or dead dog, cat, monkey (nonhuman primate animal), guinea pig, hamster, rabbit, or other such warm-blooded animal as the Secretary may determine is being used, or is intended for use, for research, testing, experimentation, or exhibition purposes, or as a pet." In this way, the Secretary of the Department of Agriculture was provided the authority to determine which animals would be covered by the Act. In 1977 the USDA promulgated regulations that specifically excluded rats, mice, and birds used for research from the definition of "animal." In 2002, the Helms amendment to the Farm Security and Rural Investment Act (also known as the 2002 Farm Bill) explicitly excluded "birds, mice of the genus *Mus*, and rats of the genus *Rattus*, bred for use in research" from the definition of "animal," thereby aligning the definitions in both the law and the regulations. However, these animals are covered under the Act for other purposes (e.g., zoos, aquaria). The rationale for Congress to accept their exclusion from the Act was based in large part on the fact that these species are covered by other federal (e.g., PHS) and private (e.g., AAALAC International) systems of oversight. As reports consistently estimate that approximately 95% of research using rats and mice is funded by the NIH, these animals are covered by the Health Research Extension Act/Public Health Service Policy (Bayne et al., 2010) and thus receive federal oversight.

3.2.2 TRANSLATING THE ANIMAL WELFARE ACT INTO ACTION

The AWA authorizes the USDA, through the Secretary of Agriculture, to develop standards, rules, regulations, and orders based on the Act. Within the USDA, the AWA is administered through the Animal and Plant Health Inspection Service (APHIS). To ensure harmonization across the federal government, all regulations proposed by the USDA must be developed in consultation and cooperation with other federal departments and agencies that govern research animals and be reviewed and to ensure fiscal feasibility, approved by the Office of Management and Budget. The USDA is required to publish any new regulations or changes in existing regulations in the *Federal Register* and allow a 60-day period during which the public may comment. The final rule on the regulations is published in the *Federal Register*, along with an effective implementation date.

The USDA regulations and standards are published as the Animal Welfare Regulations in the Code of Federal Regulations (CFR), Title 9, Animals and Animal

Products, Subchapter A, Animal Welfare. Part 1 defines the terms used (e.g., the definition of "animal" appears in this part), Part 2 provides the regulations, Part 3 specifies the standards, and Part 4 includes the rules of practice governing proceedings under the AWA. In addition, the USDA has issued an "Animal Care Policy Manual" to further clarify the intent of the Animal Welfare Regulations. The principal components of the Animal Welfare Regulations that pertain to research facilities are provided in Part 2, Subparts C and D, which are summarized later.

(a) Regulated Species

Species covered by the USDA Animal Welfare Regulations include any live or dead dog, cat, nonhuman primate, guinea pig, hamster, rabbit, aquatic mammal, or any other warm-blooded animal that is being used or is intended for use in research, teaching, testing, experimentation, exhibition, or as a pet. As noted previously, birds, rats of the genus *Rattus*, and mice of the genus *Mus* bred for use in research, teaching, or testing, and horses and farm animals, intended for use as food or fiber are used in studies to improve production and quality of food and fiber, are specifically excluded.

(b) Licensing

Any person operating or desiring to operate as a dealer, broker, exhibitor, or operator of an auction sale must be licensed by the USDA and pay an annual fee. A dealer is any person who, for compensation or profit of more than $500 per year, buys, sells, or negotiates the purchase of, delivers for transportation, or transports a regulated animal for research, teaching, testing, experimentation, or exhibition or for use as a pet or a dog for hunting, security, or breeding purposes. Traditional "brick and mortar" retail pet stores are exempted unless they sell to a research facility, exhibitor, or wholesale dealer (although the USDA has recently begun requiring federal licensing and inspections for Internet-based businesses and others that sell animals sight unseen). Dogs and cats acquired by a dealer or exhibitor must be held for 5 full days, not including the day of acquisition, after acquiring the animal. If the animal was acquired from a contract animal pound or shelter, the animal must be held for at least 10 full days. If the animal is then sold to another dealer, the subsequent dealer is required to hold the animal for a minimum of 24 h. Research facilities that obtain dogs and cats from sources other than dealers, exhibitors, and exempt persons must also hold the animals for 5 full days, not including the day of acquisition or time in transit, before the animals are used by the facility.

(c) Registration

Research facilities, intermediate handlers, and common carriers of regulated species must register with the USDA every 3 years; any revisions to the initial registration must be provided at the time of reregistration. Research facilities are defined as any institution, organization, or person that uses live animals in research, testing, or experiments; purchases or transports live animals; or receives federal funds for research, tests, or experiments. The Secretary of Agriculture may exempt facilities from registration if they do not use cats, dogs, or a substantial number of other regulated species.

(d) Institutional Animal Care and Use Committee Responsibilities

To help ensure humane experimental animal use, the 1985 AWA amendments required every animal research facility to establish an Institutional Animal Committee, subsequently designated by the USDA as an Institutional Animal Care and Use Committee (IACUC). The composition and functioning of the IACUC is covered in greater detail in a later chapter of this book.

(e) Personnel Qualifications

The 1985 AWA amendments also, for the first time, required research facilities to ensure that all scientists, research technicians, animal technicians, and other personnel involved with animal care, treatment, and use are qualified to perform their duties. This responsibility requires each institution to provide training and instruction on the humane methods of animal maintenance and experimentation, including the basic needs and the proper handling and care of each species, pre- and postprocedural care of animals, and methods of aseptic surgery. Personnel must also be instructed about research or testing methods that minimize or eliminate the use of animals or limit animal pain or distress and the utilization of information services that would help them search for alternatives. In addition, they must be informed about the methods whereby deficiencies in animal care and treatment should be reported.

(f) Information Services

To support the required training, the 1985 AWA amendments mandated the Secretary of Agriculture to establish information services at the NAL to provide (1) information pertinent to employee training; (2) methods that could prevent unintended duplication of animal experimentation as determined by the needs of the research facility; and (3) improved methods of animal experimentation that could reduce or replace animal use and minimize pain and distress, such as anesthetic and analgesic procedures. The Animal Welfare Information Center, NAL (https://awic.nal.usda.gov/awic-services), meets these requirements.

(g) Attending Veterinarian

Each research facility is required to have an attending veterinarian with training or experience in laboratory animal science and medicine who has direct or delegated program responsibility for activities involving animals at the research facility. Part-time or consulting veterinarians must provide a written program of veterinary care and perform regularly scheduled visits to the research facility. The veterinarian is authorized to ensure the provision of adequate veterinary care and to oversee the adequacy of animal care and use. Adequate veterinary care includes the availability of appropriate facilities, personnel, equipment, and services. It also includes the use of appropriate methods to prevent, control, diagnose, and treat diseases and injuries and the provision of emergency veterinary medical care. The veterinarian, through his/her role of oversight of the program of veterinary care must ensure that all animals are observed at least once daily to assess their health and well-being (which can be accomplished by someone other than the attending veterinarian but communicates directly and frequently to the attending veterinarian). The veterinarian is also responsible for providing guidance to investigators and other personnel regarding the handling, immobilization, anesthesia, analgesia, tranquilization, and euthanasia of animals. Adequate pre- and postprocedural care must be provided in accordance with current established veterinary medical and nursing practices.

(h) Records

The USDA requires that records be maintained for each IACUC meeting; each proposed activity involving animals, including any significant changes; the status of IACUC approval for each activity or change; and semiannual IACUC reports and recommendations. Every research facility must also maintain records concerning any dog or cat purchased, owned, held, transported, euthanized, or sold. These records must document the animal's source and date of acquisition, USDA-designated unique identification tag or tattoo number, species or breed, sex, date of birth or approximate age, and any distinguishing physical characteristics. The transportation, selling, or other disposition of a dog or cat must also be documented, including the name and address of the carrier (if transported) and the new owner (if sold or donated). With the exception of the source and date of acquisition, these records must accompany any shipment of dogs or cats. A health certificate signed by a licensed veterinarian must accompany all shipments of dogs, cats, and nonhuman primates. Records that relate directly to activities approved by the IACUC must be maintained for the duration of the activity and for an additional 3 years after completion of the activity. A copy of all other records and reports must be maintained for 3 years and shall be available for inspection and copying by authorized APHIS or federal funding agency representatives.

Each research facility must submit an annual report to the USDA on or before December 1 of each calendar year to provide information relevant to the immediately preceding fiscal year (October 1–September 30). The report must assure (1) that professionally acceptable standards governing the care, treatment, and use of animals were followed; (2) that each principal investigator has considered alternatives to painful procedures; and (3) that the facility is adhering to the USDA standards and regulations, unless the IACUC has approved exceptions specified and explained by the principal investigator. A summary of any exceptions, including a brief explanation and the species and number of animals affected, must be attached to the annual report. In addition, the report must state the location of all facilities where animals were housed or used in actual research, testing, teaching, or experimentation or were held for these purposes. The common names and the numbers of animals used must be reported in one of the three categories: (1) activities involving no pain, distress, or use of pain-relieving drugs; (2) experiments, teaching, research, surgery, or tests where appropriate anesthetic, analgesic, or tranquilizing drugs were used; and (3) painful activities where the use of pain-relieving agents would have adversely affected the procedures, results, or interpretation of the activity. An explanation of the animal procedure(s) conducted in the third category must be attached to the annual report. In addition, the number of animals being bred, conditioned, or held for use, but not yet used, must be listed. The USDA compiles the information contained in the reports from all registered research facilities and submits an annual summary to Congress.

3.2.3 IMPLEMENTING THE DETAILS OF RESEARCH ANIMAL CARE AND USE—THE USDA STANDARDS

The USDA "Standards" are described in Part 3 of the Animal Welfare Regulations (CFR, 1999a). Dealers, exhibitors, and research facilities are required to meet minimal housing, operating, animal health, husbandry, and transportation standards. These include feeding, watering, sanitation, lighting, ventilation, shelter from extremes of weather and temperatures, adequate veterinary care, and separation by species where the Secretary of Agriculture finds it necessary for the humane handling, care, or treatment of animals. Specific standards, including minimal enclosure space requirements, are provided for dogs, cats, guinea pigs, hamsters, rabbits, nonhuman primates, and marine mammals. The specifications are similar for all species, except marine mammals, and are more detailed for dogs, cats, and nonhuman primates. General standards are also provided for other warm-blooded species, including farm animals used for biomedical research purposes or for testing and production of biologicals for humans or nonagricultural or nonproduction animals. It is beyond the scope of this chapter to provide detailed information regarding these standards. However, several of the most notable aspects, as required by the 1985 amendments to the AWA, are summarized in the following sections.

(a) Canine Opportunity for Exercise

Dogs housed in the same primary enclosure must be compatible. Dealers, exhibitors, and research facilities must develop, document, and follow an appropriate plan, approved by the attending veterinarian, to provide dogs over 12 weeks of age with the opportunity for exercise. This rule does not apply to individually housed dogs provided with at least twice the minimum floor space required or to dogs that are group-housed in floor space that meets the minimum space standards for each dog. Bitches with litters and incompatible, aggressive, or vicious dogs are also exempted. The attending veterinarian may also exempt dogs from this program if participation would adversely affect the dog's health or well-being. Such exemptions made by the attending veterinarian must be documented and reviewed at least every 30 days by the veterinarian, unless the condition is permanent. The IACUC may also approve exemptions if the principal investigator determines that it is inappropriate for certain dogs to exercise or be group-housed. This exception must be reviewed annually by the IACUC. Records of these exemptions must be maintained and made available to the USDA or federal funding agency upon request. If a dog is housed without sensory contact with another dog, it must be provided with positive contact with humans at least daily.

(b) Psychological Well-Being of Nonhuman Primates

Dealers, exhibitors, and research facilities must develop, document, and follow an appropriate plan for environmental enhancement adequate to promote the psychological well-being of nonhuman primates. The plan must be in accordance with currently accepted professional standards as cited in appropriate professional journals or reference guides and as directed by the attending veterinarian. At a minimum, the plan must address the social needs of nonhuman primate species known to exist in social groups in nature. Individual animals that are vicious, over-aggressive, or debilitated should be individually housed. Nonhuman primates that are suspected of having a contagious disease must be isolated from healthy animals in the colony as determined by the attending veterinarian. Group-housed nonhuman primates must be determined to be compatible in accordance with generally accepted professional practices and actual observations, as directed by the attending veterinarian. Individually housed nonhuman primates must be able to see and hear members of their own or compatible species unless the attending veterinarian determines that this arrangement would endanger their health, safety, or well-being.

Primary enclosures must be enriched by providing means of expressing non-injurious species-typical behavior. Environmental enrichment devices may include perches, swings, mirrors, manipulanda, and foraging or task-oriented feeding methods. Interaction with familiar and knowledgeable personnel is recommended, provided it is consistent with safety precautions. Special attention is required for infant and young juvenile nonhuman primates, those that exhibit signs of psychological distress, those entered in IACUC-approved research protocols that require restricted activity, and individually housed nonhuman primates without sensory contact with nonhuman primates of their own or compatible species. Great apes weighing more than 110 lb must be provided additional opportunities to express species-typical behavior. If a nonhuman primate must be maintained in a restraint device for an IACUC-approved protocol, such restraint must be for the minimum period possible. If the protocol requires more than 12 h of continuous restraint, the nonhuman primate must be provided the daily opportunity for at least one continuous hour of unrestrained activity, unless the IACUC approves an exception. Such an exception must be reviewed at least annually. The attending veterinarian may also exempt an individual nonhuman primate from participation in the environmental enhancement plan in consideration of its well-being. However, such an exemption must be documented and reviewed by the attending veterinarian every 30 days. All exemptions must be available for review by the USDA and federal funding agencies on request and reported in the annual report to the USDA.

3.2.4 ENSURING COMPLIANCE WITH THE REGULATIONS

The USDA is also charged with enforcement of the Animal Welfare Regulations. The Animal Care section of the USDA's APHIS is responsible for ensuring compliance of transportation, sale, and handling of animals used in laboratory research. The Act requires the USDA to inspect each research facility at least once each year and, in the case of deficiencies or deviations from the standards promulgated under the Act, to conduct follow-up inspections as necessary until all deficiencies or deviations are corrected. APHIS enforces the AWA through unannounced inspections. Federal research facilities are exempted from this provision. Investigations may also be conducted by APHIS' Investigative and Enforcement Service personnel as the result of alleged violations of the AWA in response to public or internal complaints. Animal Care uses a risk-based inspection system which allows inspectors to conduct more frequent and in-depth inspections at problem facilities and fewer (though no less than annual) inspections at facilities that are consistently in compliance. Each research facility is required to permit APHIS officials to enter its place of business; to examine and make copies of the required records; to inspect the facilities, property, and animals; and to document, by taking photographs and other means, conditions, and areas of noncompliance. Should an AWA violation be confirmed, the USDA can take a variety of actions. Many infractions are resolved with an official notice of warning or a stipulation offer, which allow the institution to pay a penalty in lieu of formal administrative proceedings. Should a serious or chronic violation be identified, a department-level review would occur with the issuance of a formal administrative complaint, which may be resolved by license suspensions, cease-and-desist orders, civil penalties, or a combination of these penalties through administrative procedures (https://www.aphis.usda.gov/aphis/ourfocus/animalwelfare/enforcementactions/). The Animal Care Information System Search Tool (https://acissearch.aphis.usda.gov/LPASearch/faces/Warning.jspx) allows for online searches of inspection reports, including animal inventories, inspection report citations, and the number of covered species of animals used in research.

3.3 PUBLIC HEALTH SERVICE POLICY

The Health Research Extension Act of 1985 (P.L. 99−158), Section 495, Animals in Research, mandates the Secretary of Health and Human Services, acting through the Director of the NIH, establish guidelines for the proper care and treatment of animals used in biomedical and behavioral research. Any institution receiving support through the PHS for animal research, training, biological testing, or animal-related activities, must provide extensive written assurance of their compliance with the PHS Policy on Humane Care and Use of Laboratory Animals (PHS Policy; OLAW, 2002). The Policy applies to all PHS-conducted or supported activities involving animals regardless of where they are conducted. Most federally supported animal-based biomedical research in the United States is funded through the Department of Health and Human Services including the NIH; CDC; FDA; and EPA.

The PHS Policy was first implemented in 1973, and then revised in 1979, 1986, and 2002. The PHS Policy stipulates that institutions must have an Animal Welfare Assurance Statement (using the outline at http://grants.nih.gov/grants/olaw/sampledoc/assur.htm) approved by the Office of Laboratory Animal Welfare (OLAW), NIH, and an IACUC that is responsible for reviewing proposed projects, evaluating the animal care and use program, and inspecting facilities. The IACUC must also maintain records of its activities and report at least annually to OLAW.

The Assurance Statement commits the institution to follow the US Government Principles for the Utilization and Care of Vertebrate Animals Used in Testing, Research, and Training (IRAC, 1985; see Table 3.1) and the *Guide for the Care and Use of Laboratory Animals* (*Guide*; NRC, 2011b). Because the PHS Policy's standards for animal care and use are based on the *Guide*, the PHS Policy covers all vertebrate animals, including traditional laboratory animals, farm animals, wildlife, and aquatic animals used in research, testing, or teaching and thus is broader than the reach of the USDA Animal Welfare Regulations in this regard. The *Guide* emphasizes performance standards, which are less prescriptive and more context-specific than rigid engineering standards. It also encourages the application of professional judgment and highlights the importance of the institutional official, IACUC, and attending veterinarian to function as a team in the oversight of the animal care and use program. Recommendations in the *Guide* are based on published data, scientific principles, expert opinion, and experience with methods and practices that are consistent with high-quality, technically and scientifically appropriate, humane animal care and use. The *Guide* provides recommendations for occupational health and safety programs. Numerous relevant references are provided that provide further background to the recommendations of the *Guide*.

In addition to stating a commitment to animal welfare, the Assurance Statement must (1) designate clear lines of authority and responsibility for institutional oversight of the research, inclusive of a designated "Institutional Official" who is ultimately responsible for the animal care and use program; (2) identify a qualified veterinarian who is involved in the program; (3) provide a description of the occupational health and safety program for relevant personnel in the program; (4) describe mandated training; and (5) provide a description of the facility. The institution must also indicate whether the animal care and use program is reviewed by a third party, such as AAALAC International, or that the program and facilities are reviewed only by internal systems of the institution. Institutions in this latter category must provide a copy of their most recent semiannual report with the Assurance Statement. The Assurance Statement is renegotiated with OLAW every 5 years. OLAW can approve, disapprove, restrict, or withdraw approval of the Assurance Statement.

PHS funding agencies, such as the NIH, may not make an award for an activity involving live vertebrate animals unless the prospective awardee institution and all

Table 3.1 US Government Principles for the Utilization and Care of Vertebrate Animals Used in Testing, Research, and Training

I	The transportation, care, and use of animals should be in accordance with the Animal Welfare Act (*U.S. Code*, Vol. 7, Secs. 213b et seq.) and other applicable federal laws, guidelines, and policies.
II	Procedures involving animals should be designed and performed with due consideration of their relevance to human or animal health, the advancement of knowledge, or the good of society.
III	The animals selected for a procedure should be of an appropriate species and quality and the minimum number required to obtain valid results. Methods such as mathematical models, computer simulation, and in vitro biological systems should be considered.
IV	Proper use of animals, including the avoidance or minimization of discomfort, distress, and pain when consistent with sound scientific practices, is imperative. Unless the contrary is established, investigators should consider that procedures that cause pain or distress in human beings may cause pain or distress in other animals.
V	Procedures with animals, that may cause more than momentary or slight pain or distress, should be performed with appropriate sedation, analgesia, or anesthesia. Surgical or other painful procedures should not be performed on unanesthetized animals paralyzed by chemical agents.
VI	Animals that would otherwise suffer severe or chronic pain or distress that cannot be relieved should be painlessly killed at the end of the procedure, or, if appropriate, during the procedure.
VII	The living conditions of animals should be appropriate for their species and contribute to their health and comfort. Normally the housing, feeding, and care of all animals used for biomedical purposes must be directed by a veterinarian or other scientist trained and experienced in the proper care, handling, and use of the species being maintained or studied. In any case, veterinary care shall be provided as indicated.
VIII	Investigators and other personnel shall be appropriately qualified and experienced for conducting procedures on living animals. Adequate arrangements shall be made for their in-service training, including the proper and humane care and use of laboratory animals.
IX	Where exceptions are required in relation to the provisions of these principles, the decisions should not rest with the investigators directly concerned but should be made, with due regard to Principle II, by an appropriate review group such as the institutional animal research committee. Such exceptions should not be made solely for the purposes of teaching or demonstration.

other institutions participating in the animal activity have an approved Assurance Statement with OLAW and provide verification that the IACUC has reviewed and approved those sections of the grant application that involve the use of animals. Applications from organizations with approved Assurance Statements must address five specific points pertaining to the use of animals:

- a detailed description of the proposed work, including species, strain, sex, age, and number of animals to be used in the proposed work;
- a justification of the use of animals, species, and number of animals;
- information on the veterinary care for the animals;
- a description of the procedures for ensuring that discomfort, distress, pain, and injury will be minimized; and
- a description of the method of euthanasia and the reason for the selection of that method, including a justification for any method that does not conform with the American Veterinary Medical Association's (AVMA) Euthanasia Guidelines.

Awardee institutions that do not comply with the standards of the *Guide*, the USDA Animal Welfare Regulations, and other standards referenced in the PHS Policy (e.g., the AVMA's Report of the Panel on Euthanasia, 2013), may have their Assurance Statement restricted, which in turn can limit access to PHS funding for research. Sustained noncompliance with the PHS Policy can result in withdrawing the approval of the Assurance Statement and cessation of all PHS funding for animal-based activities.

The awardee institution must also submit an annual report. Institutions must report any change in category status (i.e., accredited by AAALAC International or not) from that noted in the Assurance Statement. Institutions indicate the dates of their IACUC's semiannual program reviews and facility inspections and provide with the annual report copies of any "minority views" filed by IACUC members.

OLAW conducts site visits of awardee institutions both "for cause" and "not for cause." In addition, an ongoing significant mission of OLAW is the educational outreach it performs in collaboration with awardee institutions. Jointly sponsored workshops focus on information of value to Institutional Officials and IACUC's to provide appropriate oversight of animal care and use. OLAW also provides guidance through articles in journals, commentary on other articles, NIH guide notices, frequently asked questions, and a listserve.

OLAW also refers to the NIH Revitalization Act of 1993 (P.L. 103–43), "Plan for Use of Animals in Research" (http://grants.nih.gov/grants/olaw/pl103-43.pdf), which requires the Director of the NIH to prepare a plan:

1. for the NIH to conduct or support research into:
   a. methods of biomedical research and experimentation that do not require the use of animals,
   b. methods of such research and experimentation that reduce the number of animals used in such research,

    c. methods of such research and experimentation that produce less pain and distress in such animals, and

    d. methods of such research and experimentation that involve the use of marine life (other than marine mammals);

2. for establishing the validity and reliability of the method(s) described in list (1);

3. for encouraging the acceptance by the scientific community of such methods that have been found to be valid and reliable; and

4. for training scientists in the use of such methods that have been found to be valid and reliable.

The NIH Revitalization Act is one key method by which the three R's (Russell and Burch, 1959) are implemented in the US research enterprise.

3.4 FDA GOOD LABORATORY PRACTICES

The Federal Food, Drug, and Cosmetic Act requires the FDA to ensure proper procedures for the care and use of laboratory animals, as implemented by the Good Laboratory Practice (GLP) regulations (21 CFR, Part 58) that became effective in June 1979 and were most recently amended in 2002 (for a summary, see http://www.nabranimallaw.org/Research_Animal_Protection/Good_Laboratory_Practice_for_Nonclinical_Laboratory_Studies/). The regulations establish basic standards for conducting and reporting nonclinical safety testing and are intended to assure the quality and integrity of safety data submitted to the FDA in support of an application for a research or marketing permit. Such permits are required for human and animal drugs, human biological products, medical devices, diagnostic products, food and color additives, and electronic medical products. Basic research studies, clinical or field trials in animals, and human subject trials are not covered by the GLP regulations.

    Institutions seeking FDA approval of their products must establish written protocols and SOPs; provide adequate facilities, equipment, and animal care; properly identify test substances; and accurately record observations and report results for preclinical studies (i.e., research conducted before clinical trials in humans). The FDA uses documentation of adherence to the written protocols and SOPs in judging the acceptability of safety data submitted in support of marketing or clinical research permits. Every study conducted under GLP regulations must have a study director who is ultimately responsible for the implementation of the protocol and conduct of the study. Each institution must also have a quality assurance unit that monitors the conduct of studies to ensure the protocol is being followed and the records are properly maintained.

    To help ensure compliance with the GLP regulations, the FDA conducts periodic and routine surveillance inspections and data audits of public, private, and government nonclinical laboratories that may be performing tests on GLP-regulated products. They may also conduct directed inspections to verify the reliability, integrity, and compliance of important or critical safety studies being reviewed in support of pending applications for product research or premarketing approval. In addition, the FDA may conduct inspections to investigate potential noncompliance issues brought to the FDA by "whistle-blowers," the news media, industry complaints, FDA reviewers, other government contacts, or other sources. Inspections of commercial laboratories are conducted without prior notification. Initial inspections of university and government laboratories are initiated only after the facility has been informed in a letter, from the Bioresearch Monitoring Program coordinator, Division of Compliance Policy, Office of Enforcement, FDA, of the intent to inspect.

The inspections include a review of the institution's organization and personnel, quality assurance unit, facilities, equipment, testing facility operations, reagents and solutions, test and control articles, protocols and conduct of nonclinical studies, records, and reports. In addition, the animal care program is evaluated to determine if the animal care and housing is adequate to preclude stress and uncontrolled influences that could alter the response of the test system to the test article. The inspection includes the animal housing room(s) and SOPs for the environment, housing, feeding, handling, and care of laboratory animals. The audit will verify that newly received animals are appropriately isolated, identified, and evaluated for health status. It will also confirm that animals of different species, or animals of the same species on different projects, are separated. Daily logs of animal health observations are randomly reviewed and records of animal treatments are evaluated to ensure they have been properly authorized and documented. In addition, documentation of cage, rack, and accessory equipment sanitation, as well as the use of appropriate bedding are assessed. Analysis of feed and water samples, collected at appropriate sources and proper frequency, and the pest control program are also reviewed. Copies of the IACUC's SOPs and meeting minutes are reviewed to verify committee operation.

A data audit is also conducted to compare the protocol and amendments, raw data, records, and specimens against the final safety assessment report. This audit is intended to substantiate that protocol requirements were met and findings were fully and accurately reported. The study methods described in the final report are compared against the protocol and SOPs to confirm that the GLP requirements were met. In addition to reviewing the procedures and methods for animal housing, identification, health observations, and treatment, the audit includes review of the handling of dead or moribund animals and necropsy, histopathology, and pathology procedures. The audit also includes a detailed review of study records and raw data. These data may include animal weight records, food consumption records, and clinical pathology analyses and ophthalmologic examinations.

Inspection reports are classified according to the findings and whether or not objectionable conditions or practices were found during the inspection. If regulatory and/or administrative actions are recommended, the FDA may hold an informal conference, conduct a reinspection, or issue a warning letter. It may also reject a nonclinical study or studies, disqualify the institution, withhold or revoke a marketing permit, or terminate a permit for preclinical studies.

3.5 OTHER LAWS, REGULATIONS, AND POLICIES

3.5.1 ENVIRONMENTAL PROTECTION AGENCY

The EPA is responsible for the registration and control of pesticides (in accordance with the Federal Insecticide, Fungicide and Rodenticide Act of 1947, P.L. 80−104) and approval of industrial chemicals (under the Toxic Substance Control Act of 1976). Evaluation of pesticides and industrial chemicals require animal testing, to assess potential health and environmental effects, conducted in accordance with the EPA's GLP standards (40 CFR, Part 160; for a summary, see http://www.nabranimallaw.org/Research_Animal_Protection/EPA_Good_Laboratory_Practice_Standards_(full_text)/). The GLPs address animal husbandry [housing (including appropriate water systems for marine and fresh water organisms), feeding, handling and care] and stipulate detailed documentation, including the establishment of and adherence to SOPs. EPA officials inspect the testing facilities, review the data, and prepare inspection reports. Noncompliance with the EPA GLPs may result in the agency refusing to consider experimental data in support of an application, disqualification of a testing facility, or criminal prosecution.

3.5.2 DEPARTMENT OF DEFENSE

The DoD developed a "Policy on Experimental Animals in Department of Defense Research" in 1961 to ensure that all research at DoD facilities involving animals was conducted in accord with certain principles of animal care. In 1995, the DoD issued a policy memorandum, "Department of Defense (DoD) Policy for Compliance with Federal Regulations and DoD Directives for the Care and Use of Laboratory Animals in DoD-Sponsored Programs" (for a summary, see http://www.nabranimallaw.org/Research_Animal_Protection/Department_of_Defense_(DoD)/), which defined animal to include rats of the genus *Rattus*, mice of the genus *Mus*, and birds, thereby expanding on the AWA definition. The policy memorandum also required all DoD facilities maintaining animals used in research, testing, or teaching to apply for accreditation by AAALAC International. In 2005, the DoD issued "The Care and Use of Laboratory Animals in DoD Programs" (http://armypubs.army.mil/epubs/pdf/r40_33.pdf). Among several other stipulations, this administrative revision prohibited the wounding of dogs, cats, and nonhuman primates for medical or surgical training and their use in advanced trauma life support training; added a policy for managing complaints concerning violations of animal care and use standards; and specified a requirement to release information to the public in an accessible database. In 2010 a DoD Instruction (3216.01), entitled "The Use of Animals in DoD Programs" (http://www.dtic.mil/whs/directives/corres/pdf/321601p.pdf), extended the scope to vertebrate animals—alive and dead. It further required that "methods other than animal use and alternatives to animal use (i.e., methods to refine, reduce, or replace the use of animals) shall be considered and used whenever possible to attain the objectives of research, development, test, and evaluation or training if such alternative methods produce scientifically or educationally valid or equivalent results." The instruction required all DoD facilities to attain and maintain accreditation by AAALAC International unless the DoD site housed animals for less than eight continuous calendar days, specifies the IACUC membership for intramural DoD facilities and describes DoD requirements for DoD-funded research at extramural sites.

3.5.3 STATE LAWS

State laws to protect animals have a long history; the first state anticruelty law was passed in 1641 in the Massachusetts Bay Colony to prevent riding or driving farm animals beyond established limits. All 50 states and the District of Columbia have enacted anticruelty laws; most states include felony provisions. The overarching goals of these laws are to protect animals from cruel treatment, require that animals have access to suitable food and water, and require that animals have shelter from extreme weather. Some state laws define "animal" and some do not. The state laws encompass a diversity of approaches to providing protection to animals. Some states have additional provisions for animals used in research, and many states prohibit the sale of pound animals into the research supply chain. The majority of states have laws that protect animal research facilities from arson, economic sabotage, and other acts of intimidation. Several states address the use of animals in K-12 classrooms, to include dissection exercises, so that the student is offered an alternative means of instruction such as videos or computer simulations. In recent years, state and federal laws have been used by private citizens or citizen groups claiming "standing to sue" on behalf of animals. The issue of "standing" has undergone a long litigation process, and a chronology of court decisions on this issue has been compiled by the National Association for Biomedical Research (http://www.nabranimallaw.org/Federal/AWA_Case_Law_(Listed_by_Subject)/).

3.5.4 OTHER FEDERAL LAWS

Because animal research can involve a variety of different species, several other federal acts, laws, and treaties have bearing on animal use. These include the US Endangered Species Act (P.L. 93–205, 1973), which restricts the research conducted on designated animals to investigations that would directly benefit the species being studied; the Marine Mammal Protection Act (P.L. 92–522, 1972), which provides authority for scientific research on marine mammals by special permit; the Convention on International Trade in Endangered Species of Wild Fauna and Flora (CITES, 1973), which requires signatory countries to obtain a permit for the import or export of certain species; the Lacey Act (1900 et seq.), which governs trade in wildlife, fish, and plants that have been illegally taken, possessed, transported, or sold; and the Migratory Bird Treaty Act (1918), which makes it unlawful to pursue, hunt, take, capture, kill, or sell listed species of birds except by permit.

3.6 FISH AND WILDLIFE SERVICE AND NIH POSITIONS ON CHIMPANZEES

After accepting a December 2011 Institute of Medicine (IOM) report, "Chimpanzees in Biomedical and Behavioral Research: Assessing the Necessity" (IOM, 2011), the NIH convened a Working Group on the Use of Chimpanzees in NIH-Supported Research to advise on the implementation of the IOM report. The Working Group's report, published in January 2013 contained 28 recommendations and was made available for public comment [Council of Councils, 2013; http://dpcpsi. nih.gov/sites/default/files/FNL_Report_WG_Chimpanzees_0.pdf]. In June 2013, the NIH announced its plans to substantially reduce the use of chimpanzees in NIH-funded research and to retire the majority of chimpanzees it owns.

Specifically, the NIH plans to:

- retain but not breed up to 50 chimpanzees for future research that meets the IOM principles and criteria (http://iom.edu/Reports/2011/Chimpanzees-in-Biomedical-and-Behavioral-Research-Assessing-the-Necessity.aspx);
- provide ethologically appropriate facilities (i.e., as would occur in their natural environment) for those chimpanzees as defined by NIH, with space requirements yet to be determined;
- establish a review panel to consider research projects proposing the use of chimpanzees with the IOM principles and criteria after projects have cleared the NIH peer review process;
- wind down research projects using NIH-owned or NIH-supported chimpanzees that do not meet the IOM principles and criteria in a way that preserves the research and minimizes the impact on the animals; and
- retire the majority of the NIH-owned chimpanzees deemed unnecessary for biomedical research to the Federal Sanctuary System contingent upon resources and space availability in the sanctuary system.

In a separate action, in June 2013 the US Fish and Wildlife Service (FWS) proposed (http://www.gpo.gov/fdsys/pkg/FR-2013-06-12/pdf/2013-14007.pdf) to modify the classification of captive chimpanzees from "threatened" to "endangered." After receipt of a petition in 2010 from a coalition of organizations, including the Humane Society of the United States, the World Wildlife Fund and the Jane Goodall Institute to list all chimpanzees as endangered based on the grounds that the use of chimpanzees in research, entertainment, and as pets further impairs the status of chimpanzees in the wild, the FWS conducted a formal review of the status of the chimpanzee under the ESA. The FWS determined that the ESA does not allow for captive-held animals to be assigned a separate legal status from their wild counterparts. With the classification of the chimpanzee as endangered, certain activities would require a permit, including import and export of chimpanzees into and out of the United States, "take" (defined by the ESA as harm, harass, kill, injure, etc.) within the United States, and interstate and foreign commerce. Permits could be issued for scientific purposes or to enhance the propagation or survival of the animals.

3.7 PROFESSIONAL AND SCIENTIFIC ASSOCIATIONS

Many scientific and professional associations have adopted position statements regarding the care and use of laboratory animals. Several of these also provide resources that have made a significant impact on the generally accepted practices for animal facilities.

3.7.1 INSTITUTE FOR LABORATORY ANIMAL RESEARCH

The NAS is a nongovernmental, nonprofit organization chartered by Congress in 1863 to "investigate, examine, experiment, and report upon any subject of science" (http://www.nas.edu/about/whoweare/index.html). The ILAR is a component of the Division on Earth and Life Studies, one of the six subject area divisions in the NRC. The NRC is operated jointly by the NAS and the National Academy of Engineering, and it is the organizational unit within the National Academies that conducts most policy studies at the request of the federal government.

ILAR is advised by a council of experts in laboratory animal medicine, zoology, genetics, medicine, ethics, and related biomedical sciences. The council provides direction for ILAR's programs. Many of ILAR's reports provide a framework for governmental and institutional animal welfare policies. The most widely distributed publication from ILAR is the *Guide for the Care and Use of Laboratory Animals* (the most recent revision being the eighth edition, NRC, 2011b), which is recognized by the PHS and AAALAC International as a standard reference on laboratory animal care and use programs.

ILAR has published several other standard references that are used to establish and maintain optimal animal care and use programs: *The Psychological Well-being of Nonhuman Primates* (1998); *Recognition and Alleviation of Distress in Laboratory Animals* (2008) and its companion report *Recognition and Alleviation of Pain in Laboratory Animals* (2009a); *Scientific and Humane Issues in the Use of Random Source Dogs and Cats in Research* (2009b); *Guidance for the Description of Animal Research in Scientific Publications* (2011c); and *Animal Models for Assessing Countermeasures to Bioterrorism Agents* (2011a).

3.7.2 ASSOCIATION FOR ASSESSMENT AND ACCREDITATION OF LABORATORY ANIMAL CARE INTERNATIONAL

The AAALAC International is a private, nonprofit organization that promotes the humane treatment of animals in science through a voluntary accreditation program, a program status evaluation service, and educational programs. As of 2015, AAALAC International accredits more than 900 programs in 40 countries. In its assessments of animal care and use programs, AAALAC relies on three primary standards: the *Guide* (NRC, 2011b); the *Guide for the Care and Use of Agricultural Animals in Research and Teaching* (*Ag Guide*; FASS, 2010); and the *European Convention for the Protection of Vertebrate Animals Used for Experimental and Other Scientific Purposes* (ETS 123; Council of Europe, 1986). AAALAC International overlays these standards on the laws and regulations for the country in which the institution seeking accreditation is located and uses other specialty publications to supplement information about specific procedures or techniques related to the care and use of laboratory animals, designated as reference resources (http://www.aaalac.org/accreditation/resources.cfm). AAALAC International has a cooperation agreement with the World Organisation for Animal Health (OIE) to use Chapter 7.8 of the OIE's Terrestrial Animal Health Code (OIE, 2011; http://web.oie.int/eng/normes/mcode/en_chapitre_1.7.8.htm), "Use of Animals in Research and Education" (an AAALAC International reference resource) in its assessments of programs worldwide, with particular emphasis on countries that do not have their own legislative framework for ensuring the welfare of animals used in research, testing, or teaching. As noted previously, AAALAC International accreditation is formally recognized by the PHS.

3.7.3 AMERICAN COLLEGE OF LABORATORY ANIMAL MEDICINE

The Report of the American College of Laboratory Animal Medicine on Adequate Veterinary Care in Research, Testing, and Teaching (ACLAM, 1996) describes a program of adequate veterinary care, including (1) disease detection and surveillance, prevention, diagnosis, treatment, and resolution; (2) providing guidance on anesthetics, analgesics, tranquilizer drugs, and methods of euthanasia; (3) the review and approval of all preoperative, surgical, and postoperative procedures; (4) the promotion and monitoring of an animal's well-being before, during, and after its use; and (5) involvement in the review and approval of all animal care and use at the institution. This report is used by AAALAC International as a reference standard in its assessments of animal care and use programs.

3.7.4 AMERICAN VETERINARY MEDICAL ASSOCIATION

The AVMA has published guidelines on humane euthanasia of a different species of animals in numerous circumstances. The most recent edition was published in 2013 (AVMA, 2013), and it includes guidance on human euthanasia of animals used in research (https://www.avma.org/KB/Policies/Documents/euthanasia.pdf). The AVMA has issued a policy statement that the tenets of the three R's (Russell and Burch, 1959) should serve as overarching principles when animals are used in research, testing, or teaching.

3.7.5 FEDERATION OF ANIMAL SCIENCE SOCIETIES

To help ensure the ethical and humane treatment of farm animals used in agricultural research or teaching, the agricultural community published the *Ag Guide*. The first edition, published in 1988, was revised in 1999 and again in 2010 by the Federation of Animal Science Societies (FASS, 2010). The third edition of the *Ag Guide* is based on the premise that the housing and management of farm animals do not necessarily change because of the objectives of the research or teaching activity. It includes broad guidelines for institutional policies; agricultural animal health care; husbandry, housing and biosecurity; environmental enrichment; and animal handling and transport. Common agricultural animal species used in research and teaching are handled in more detail in species-specific chapters. The *Ag Guide* is used by AAALAC International for relevant program assessment and accreditation purposes; however, OLAW has placed some restrictions on the breadth of applicability of the *Ag Guide* for PHS-assured institutions if the institution's PHS assurance encompasses all animal research activities at the institution (http://grants.nih.gov/grants/olaw/faqs.htm).

3.7.6 OTHER PROFESSIONAL ORGANIZATIONS

Several other professional organizations have published guideline documents that either address a particular range of species used in research or fill a gap in existing guidelines. Examples include the Ornithological Council ["Guidelines to the Use of Wild Birds in Research" (Fair et al., 2010)], the American Society of Mammalogists ["Guidelines of the American Society of Mammalogists for the use of wild mammals in research" (Sikes et al., 2011)], the American Society of Ichthyologists and Herpetologists ["Guidelines for the Use of Live Amphibians and Reptiles in Field and Laboratory Research" (Beaupre et al., 2004)], as well as the Association of Primate Veterinarians ("NHP Food Restriction Guidelines" and "Social Housing Guidelines"), to name just a few. Such taxon-specific guidelines can provide a useful supplement to the more broadly based guidelines that underpin a high-quality animal use program.

3.8 INTERNATIONAL LAWS AND REGULATIONS

It is not the intent of this chapter to provide detailed information on the various international laws and standards governing the care and use of laboratory animals. Because the biomedical research community has become more global, there is increased interest in the harmonization of international standards for the care and use of laboratory animals (Bayne et al., 2014). As a result, the implementation of a legal and regulatory framework for the conduct of animal-based research is becoming increasingly common around the world. Two organizations, in particular, have established guidelines that have global reach.

3.8.1 THE WORLD ORGANISATION FOR ANIMAL HEALTH

The Office International des Epizooties (OIE) was founded in 1924; in 2003 it was renamed the World Organisation for Animal Health, but the historical acronym was retained. The OIE has a mission of creating a framework of international collaboration and information sharing to improve animal health and welfare. It is comprised of 178 member countries and is recognized as a reference organization for the WTO. The OIE publishes standards (predominantly on animal health and zoonoses) developed by expert groups on terrestrial and aquatic animals. Animal welfare was identified by the OIE as a priority in its 2001−05 Strategic Plan. Subsequently, the OIE convened a permanent Working Group on Animal Welfare. This working group initially focused on standards relating to the long distance transport of animals and to the killing of animals for both human consumption and disease control purposes. More recently, laboratory animal welfare was identified as an area of interest. As a result, an ad hoc committee of experts in laboratory animal welfare was convened with representation from different geographic regions of the world (United States, Canada, Europe, Africa, Asia, and South America) to develop baseline standards on laboratory animal care and use. By incorporating these standards in the Terrestrial Animal Health Code (OIE, 2011), the governments of the member countries have a responsibility to ensure these standards are reflected in their regulatory frameworks. Standards for laboratory animals appear in Chapter 7.8, "Use of Animals in Research and Education" (http://web.oie.int/eng/normes/mcode/en_chapitre_1.7.8.htm).

3.8.2 COUNCIL FOR INTERNATIONAL ORGANIZATIONS OF MEDICAL SCIENCES

Council for International Organizations of Medical Sciences (CIOMS) is an international, nongovernmental, nonprofit organization established jointly by the WHO and the UNESCO in 1949. It is comprised of 48 international member organizations representing biomedical disciplines and 18 national members representing national academies of sciences and medical research councils. In the laboratory animal medicine and science community, CIOMS is primarily known for its "International Guiding Principles for Biomedical Research Involving Animals" which were first promulgated in 1985 (Table 3.2). These 11 principles served as the basis for "US Government Principles for the Utilization and Care of Vertebrate Animals Used in Testing, Research, and Training." The guiding principles are designed to assist ethics committees, animal care committees, organizations, societies, and countries in developing programs for the humane care and use of animals in research and education, especially those entities operating without federal or national regulations.

Recently, in collaboration with the International Council for Laboratory Animal Science (ICLAS), the guiding principles have been updated to reflect contemporary opinion on the proper use of animals in research (http://grants.nih.gov/grants/olaw/Guiding_Principles_2012.pdf). Like the original 1985 guiding principles, the

Table 3.2 Council for International Organizations of Medical Sciences—International Council for Laboratory Animal Science International Guiding Principles for Biomedical Research (2012)

I	The advancement of scientific knowledge is important for improvement of human and animal health and welfare, conservation of the environment, and the good of society. Animals play a vital role in these scientific activities and good animal welfare is integral to achieving scientific and educational goals. Decisions regarding the welfare, care, and use of animals should be guided by scientific knowledge and professional judgment, reflect ethical and societal values, and consider the potential benefits and their impact on the well-being of the animals involved.
II	The use of animals for scientific and/or educational purposes is a privilege that carries with it moral obligations and responsibilities for institutions and individuals to ensure the welfare of these animals to the greatest extent possible. This is best achieved in an institution with a culture of care and conscience in which individuals work with animals willingly, deliberately, and consistently act in an ethical, humane, and compliant way. Institutions and individuals using animals have an obligation to demonstrate respect for animals, to be responsible and accountable for their decisions and actions pertaining to animal welfare, care and use, and to ensure that the highest standards of scientific integrity prevail.
III	Animals should be used only when necessary and only when their use is scientifically and ethically justified. The principles of the three R's—replacement, reduction and refinement—should be incorporated into the design and conduct of scientific and/or educational activities that involve animals. Scientifically sound results and avoidance of unnecessary duplication of animal-based activities are achieved through study and understanding of the scientific literature and proper experimental design. When no alternative methods, such as mathematical models, computer simulation, in vitro biological systems, or other nonanimal (adjunct) approaches, are available to replace the use of live animals, the minimum number of animals should be used to achieve the scientific or educational goals. Cost and convenience must not take precedence over these principles.
IV	Animals selected for the activity should be suitable for the purpose and of an appropriate species and genetic background to ensure scientific validity and reproducibility. The nutritional, microbiological, and general health status as well as the physiological and behavioral characteristics of the animals should be appropriate to the planned use as determined by scientific and veterinary medical experts and/or the scientific literature.
V	The health and welfare of animals should be the primary considerations in decisions regarding the program of veterinary medical care to include animal acquisition and/or production, transportation, husbandry and management, housing, restraint, and final disposition of animals, whether euthanasia, rehoming, or release. Measures should be taken to ensure that the animals' environment and management are appropriate for the species and contribute to the animals' well-being.
VI	The welfare, care, and use of animals should be under the supervision of a veterinarian or scientist trained and experienced in the health, welfare, proper handling, and use of the species being maintained or studied. The individual or team responsible for animal welfare, care, and use should be involved in the development and maintenance of all aspects of the

Table 3.2 Council for International Organizations of Medical Sciences—International Council for Laboratory Animal Science International Guiding Principles for Biomedical Research (2012)—cont'd

	program. Animal health and welfare should be continuously monitored and assessed with measures to ensure that indicators of potential suffering are promptly detected and managed. Appropriate veterinary care should always be available and provided as necessary by a veterinarian.
VII	Investigators should assume that procedures that would cause pain or distress in human beings can cause pain or distress in animals, unless there is evidence to the contrary. Thus, there is a moral imperative to prevent or minimize stress, distress, discomfort, and pain in animals, consistent with sound scientific or veterinary medical practice. Taking into account the research and educational goals, more than momentary or minimal pain and/or distress in animals should be managed and mitigated by refinement of experimental techniques and/or appropriate sedation, analgesia, anesthesia, noon-pharmacological interventions, and/or other palliative measures developed in consultation with a qualified veterinarian or scientist. Surgical or other painful procedures should not be performed on unanesthetized animals.
VIII	Endpoints and timely interventions should be established for both humane and experimental reasons. Humane endpoints and/or interventions should be established before animal use begins, should be assessed throughout the course of the study, and should be applied as early as possible to prevent, ameliorate, or minimize unnecessary and/or unintended pain and/or distress. Animals that would otherwise suffer severe or chronic pain, distress, or discomfort that cannot be relieved and is not part of the experimental design should be removed from the study and/or euthanized using a procedure appropriate for the species and condition of the animal.
IX	It is the responsibility of the institution to ensure that personnel responsible for the welfare, care, and use of animals are appropriately qualified and competent through training and experience for the procedures they perform. Adequate opportunities should be provided for ongoing training and education in the humane and responsible treatment of animals. Institutions also are responsible for supervision of personnel to ensure proficiency and the use of appropriate procedures.
X	While implementation of these principles may vary from country to country according to cultural, economic, religious, and social factors, a system of animal use oversight that verifies commitment to the principles should be implemented in each country. This system should include a mechanism for authorization (such as licensing or registering of institutions, scientist, and/or projects) and oversight which may be assessed at the institutional, regional, and/or national level. The oversight framework should encompass both ethical review of animal use as well as considerations related to animal welfare and care. It should promote a harm—benefit analysis for animal use, balancing the benefits derived from the research or educational activity with the potential for pain and/or distress experienced by the animal. Accurate records should be maintained to document a system of sound program management, research oversight, and adequate veterinary medical care.

revised principles (CIOMS-ICLAS, 2012) are intended to be used by the international scientific community to guide institutions in the responsible use of vertebrate animals in scientific and/or educational activities. OLAW has issued a notification that PHS-assured institutions outside the United States are required to implement the revised guiding principles (http://grants.nih.gov/grants/guide/notice-files/NOT-OD-13-096.html).

3.8.3 INTERNATIONAL AIR TRANSPORT ASSOCIATION

The International Air Transport Association (IATA) is the trade association for airlines based around the world and guides the formation of positions on industry and public policy issues. IATA coordinates the Live Animals and Perishables Board (LAPB). The objectives of the LAPB are as follows:

- The adoption of regulations for the acceptance, handling, and loading of live animals in air transport.
- The promotion of public awareness and government acceptance of the Live Animals Regulations (LAR).
- Providing for an open forum for member airlines to exchange and develop information specific to the transport of live animals and perishables.
- Promoting an open dialog with civil aviation authorities and shipping industry.

IATA publishes the LAR (http://www.iata.org/publications/pages/live-animals.aspx), the global standard for transporting animals by commercial airlines. The LAR addresses the container specification and other transportation requirements for numerous species. IATA has a cooperation agreement with the OIE to enhance veterinary research into animal health during air transport, the development and revision of international standards for air transport of live animals and perishable goods such as biological samples, as well as the technical requirements for their international transport.

3.9 CONCLUSIONS

The United States system of oversight of the use of animals in research, testing, and teaching is a matrix of laws, regulations, policies, and guidelines that have both overlapping and nonoverlapping components. The primary federal law, the AWA, is buttressed by requirements from grant funding agencies, nonprofit accreditation, state laws, and professional society guidance documents. This composite has been viewed with mixed results. The scientific community has indicated that the plethora of different systems of oversight results in regulatory burden and has called for efficiencies to be identified and adopted. Alternatively, opponents to the use of animals in research, testing, and teaching highlight the flaws in this aggregate approach. Indeed, the system of oversight continues to evolve, primarily in response to new scientific information and societal opinion. In this way, the legal framework is modified and thus may appear to be a moving target to some constituents and inadequately responsive to other constituents. This "tension" is not unique to the subject of the use of animals in research and should be viewed as an opportunity to reevaluate the reasons and methods animals are used.

REFERENCES

American College of Laboratory Animal Medicine (ACLAM), 1996. Report of the American College of Laboratory Animal Medicine on Adequate Veterinary Care in Research, Testing and Teaching.

American Veterinary Medical Association, 2013. AVMA Guidelines for the Euthanasia of Animals: 2013 Edition. https://www.avma.org/KB/Policies/Pages/Euthanasia-Guidelines. aspx?utm_source=prettyurl&utm_medium=web&utm_campaign=redirect&utm_ term=issues-animal_welfare-euthanasia-pdf.

Animal Welfare Act (P.L. 91-579), 1970. http://awic.nal.usda.gov/public-law-91-579-animal-welfare-act-amendments-1970.

Animal Welfare Act Amendments of 1976 (P.L. 94-279), 1976. http://awic.nal.usda.gov/public-law-94-279-animal-welfare-act-amendments-1976.

Association of Primate Veterinarians. Food Restriction Guidelines for Nonhuman Primates in Biomedical Research. http://www.primatevets.org/Content/files/Public/education/NHPFoodRestrictionGuidelines.pdf.

Association of Primate Veterinarians. Social Guidelines for Nonhuman Primates in Biomedical Research. http://www.primatevets.org/Content/files/Public/education/APV%20Social%20Housing%20Guidelines%20final.pdf.

Bayne, K., Bayvel, A.C.D., Williams, V., 2014. Laboratory animal welfare: international issues. In: Bayne, K., Turner, P. (Eds.), Laboratory Animal Welfare. Elsevier, New York, pp. 55–76.

Bayne, K., Morris, T., France, M., 2010. Legislation and codes of practice: a global overview. In: Kirkwood, J., Hubrecht, R. (Eds.), Eighth Edition of the UFAW Handbook on the Care and Management of Laboratory Animals and Other Animals Used in Scientific Procedures. Blackwell Publishing, Oxford, UK, pp. 107–123.

Beaupre, S.J., Jacobson, E.R., Lillywhite, H.B., Zamudio, K., 2004. Guidelines for the Use of Live Amphibians and Reptiles in Field and Laboratory Research. American Society of Ichthyologists and Herpetologists. http://www.asih.org/files/hacc-final.pdf.

CFR, 1999a. Title 9: Animals and Animal Products; Chap. 1: Animal and Plant Health Inspection Service, Department of Agriculture; Subchap. A: Animal Welfare; Parts 1, 2, 3, and 4. Office of the Federal Register, Washington, DC.

CFR, rev. 2003, 1999b. Title 40: Protection of the Environment; Chap. 1: Environmental Protection Agency; Subchap. E: Pesticide Programs; Part 160: Good Laboratory Practice Standards. Office of the Federal Register, Washington, DC. http://www.gpo.gov/fdsys/pkg/CFR-2011-title40-vol24/xml/CFR-2011-title40-vol24-part160.xml.

CFR, rev, 2002. Title 21: Food and Drugs; Chap. 1: Food and Drug Administration, Department of Health and Human Services; Subchap. A: General; Part 58: Good Laboratory Practice for Nonclinical Laboratory Studies. Office of the Federal Register, Washington, DC. http://cfr.regstoday.com/21cfr58.aspx.

Convention on International Trade in Endangered Species of Wild Fauna and Flora (CITES), 1973. http://www.cites.org/eng/disc/what.php.

Council of Councils Working Group on the Use of Chimpanzees in NIH-Supported Research, 2013. http://dpcpsi.nih.gov/sites/default/files/FNL_Report_WG_Chimpanzees_0.pdf.

Council of Europe, 1986. European Convention for the Protection of Vertebrate Animals Used for Experimental and Other Scientific Purposes (ETS 123), Strasbourg, France. http://conventions.coe.int/Treaty/en/Treaties/html/123.htm.

Council of International Medical Science Organizations (CIOMS)- International Council for Laboratory Animal Science (ICLAS), 2012. International Guiding Principles for Biomedical Research Involving Animals. http://iclas.org/wp-content/uploads/2013/03/CIOMS-ICLAS-Principles-Final1.pdf.

Endangered Species Act of 1973 (P.L. 93-205). http://www.gpo.gov/fdsys/pkg/STATUTE-87/pdf/STATUTE-87-Pg884.pdf.

Fair, J.M., Paul, E., Jones, J. (Eds.), 2010. Guidelines to the Use of Wild Birds in Research, third ed. Ornithological Council. www.nmnh.si.edu/BIRDNET/guide/index.html.

Farm Security and Rural Investment Act of 2002 (P.L. 107-171), 2002. Title X, Miscellaneous Subtitle D. Animal Welfare. http://awic.nal.usda.gov/public-law-107-171-farm-security-and-rural-investment-act-2002.

Federation of Animal Science Societies, 2010. Guide for the Care and Use of Agricultural Animals in Research and Teaching, third ed. Champaign, IL. http://www.fass.org/docs/agguide3rd/Ag_Guide_3rd_ed.pdf.

Food, Agriculture, Conservation, and Trade Act of 1990 (P.L. 101-624), 1990. Section 2503-Protection of Pets. http://awic.nal.usda.gov/public-law-101-624-food-agriculture-conservation-and-trade-act-1990-section-2503-protection-pets.

Food Security Act (P.L. 99-198, U.S. Farm Bill of 1985), Subtitle F, Animal Welfare (Improved Standards for Laboratory Animals Act), 1985. https://awic.nal.usda.gov/public-law-99-198-food-security-act-1985-subtitle-f-animal-welfare.

Federal Insecticide, Fungicide and Rodenticide Act of 1947 (P.L. 80-104), 1947. http://www.epa.gov/agriculture/lfra.html.

Health Research Extension Act of 1985 (P.L. 99-158), Section 495, Animals in Research, 1985. http://grants.nih.gov/grants/olaw/references/hrea1985.htm.

Institute of Medicine (IOM), 2011. Chimpanzees in Biomedical and Behavioral Research: Assessing the Necessity. National Academies Press, Washington, DC.

Interagency Research Animal Committee (IRAC), May 20, 1985. U.S. Government principles for utilization and care of vertebrate animals used in testing, research, and training. Federal Register.

Laboratory Animal Welfare Act of 1966 (P.L. 89-544), 1966. U.S. Code, Vol. 7, Sections. 2131−2157. http://awic.nal.usda.gov/public-law-89-544-act-august-24-1966.

Lacey Act of 1900. http://www.fws.gov/international/laws-treaties-agreements/us-conservation-laws/lacey-act.html.

Marine Mammal Protection Act (P.L. 92-522), 1972. http://www.gpo.gov/fdsys/pkg/STATUTE-86/pdf/STATUTE-86-Pg1027.pdf.

Migratory Bird Treaty Act, 1918. http://www.fws.gov/laws/lawsdigest/migtrea.html.

National Institutes of Health, June 20, 1991. Silver Spring Monkeys. NIH Fact Sheet.

National Institutes of Health Revitalization Act of 1993 (P.L. 103-43), Plan for Use of Animals in Research, 1993. http://grants.nih.gov/grants/olaw/pl103-43.pdf.

National Research Council (NRC), 1998. The Psychological Well-Being of Nonhuman Primates. National Academies Press, Washington, DC.

NRC, 2008. Recognition and Alleviation of Distress in Laboratory Animals. National Academies Press, Washington, DC.

NRC, 2009a. Recognition and Alleviation of Pain in Laboratory Animals. National Academies Press, Washington, DC.

NRC, 2009b. Scientific and Humane Issues in the Use of Random Source Dogs and Cats in Research. National Academies Press, Washington, DC.

NRC, 2011a. Animal Models for Assessing Countermeasures to Bioterrorism Agents. National Academies Press, Washington, DC.

NRC, 2011b. Guide for the Care and Use of Laboratory Animals. National Academies Press, Washington, DC.

NRC, 2011c. Guidance for the Description of Animal Research in Scientific Publications. National Academies Press, Washington, DC.

Office of Laboratory Animal Welfare (OLAW), 2002. Public Health Service Policy on Humane Care and Use of Laboratory Animals. National Institutes of Health, Department of Health and Human Services, Bethesda, Maryland.

OIE (World Organisation for Animal Health), 2011. Terrestrial Animal Health Code. http://web.oie.int/eng/normes/mcode/en_chapitre_1.7.8.htm.

Russell, W.M.S., Burch, R.L., 1959. The Principles of Humane Experimental Technique. London (UK): Methuen. Available at: http://altweb.jhsph.edu/pubs/books/humane_exp/het-toc.

Sikes, R.S., Gannon, W.L., 2011. Guidelines of the American Society of Mammalogists for the use of wild mammals in research. J. Mammal. 92, 235–253.

Toxic Substances Control Act, U.S. Code, Title 15, Chapter 53, 1976. http://www.epa.gov/oecaagct/lsca.html.

U.S. Department of Defense, 1961. Policy on Experimental Animals in Department of Defense Research. Department of Defense Instruction, Washington, DC.

U.S. Department of Defense, 1995. The Use of Laboratory Animals in DoD Programs. Directive Number 3216.1. http://www.army.mil/usapa/epubs/pdf/r40_33.pdf.

U.S. Department of Defense, 2005. Army Regulation 40–33, SECNAVINST 3900.38C, AFMAN 40-401(1), DARPAINST 18, USUHSINST 3203. The Care and Use of Laboratory Animals in DoD Programs (Washington, DC).

U.S. Department of Defense, 2010. The Use of Animals in DoD Programs. No. 3216.01. Department of Defense Instruction.

Institutional Animal Care and Use Committee

4

M.A. Suckow[1], G.A. Lamberti[2]

University of Minnesota, Minneapolis, MN, United States[1]; University of Notre Dame, Notre Dame, IN, United States[2]

CHAPTER OUTLINE

4.1 PURPOSE AND FUNCTION

In the United States, early animal care and use regulations outlined specific expectations for the care of vertebrate animals, with a particular focus on housing and environment. With time, however, it became understood by regulatory and accrediting agencies that local, institutional oversight was likely the best mechanism to build a program that consistently ensures that standards are met with respect to the animal care and use program. This local oversight is achieved through the activities of the **Institutional Animal Care and Use Committee (IACUC)**.

4.2 COMMITTEE COMPOSITION

It is the job of the IACUC to ensure that vertebrate animals are used humanely and judiciously (Silverman et al., 2014). This requires procedures that not only establish sound principles and practices for animal use, but also provide a mechanism to periodically review the animal care and use program so that improvements can be made as the field develops and as dictated by researchers' practices.

The IACUC is meant to represent a variety of viewpoints so that a robust review process is ensured when projects using animals are proposed. Toward this end, the IACUC generally includes at least the following members:

Principles of Animal Research for Graduate and Undergraduate Students.
Copyright © 2017 Elsevier Inc. All rights reserved.

1. A scientist whose work has involved the use of animal subjects;
2. A veterinarian who has experience in the medical care of laboratory animals, and who has responsibility for the care and use of activities involving animals at the institution;
3. A nonscientist; that is, someone whose main activities do not involve science. Often, this role is filled by individuals such as a clergy person, lawyer, or ethicist; and
4. A member who is not affiliated with the institution in any way. This person represents the interests of the community that is apart from the institution; thus, any affiliation, such as a family member who works for the institution, a work history at the institution, or other relationship that might result in conflict of interest must be avoided.

Though a single individual may fill more than one of the roles above (e.g., a nonscientist might also be someone who does not have an affiliation with the institution and can serve as a community member), the United States Department of Agriculture (USDA) requires at least three members (United States Department of Agriculture, 2013); in contrast, the Public Health Service (PHS) requires at least five (United States Public Health Service, 2015). Thus, the number of IACUC members depends on which set of requirements is being followed. If both sets of requirements are being followed, then the more restrictive, in this case PHS, should be applied. As with any committee, the IACUC should have a Chairperson, typically someone who is knowledgeable with respect to regulations and requirements, and who is relatively senior in the institution.

The IACUC reports to someone designated as the **Institutional Official** (IO). The IO is appointed by the Chief Executive Office of the institution and bears the ultimate responsibility for the animal care and use program. The IO also directs resources to ensure compliance with regulations and standards. Because of this responsibility, it is important that the IO be of relatively high stature within the institution.

4.2.1 THE MAIN ACTIVITIES OF THE IACUC

The main activities of the IACUC are the following:

1. Review proposed activities and projects that will use vertebrate animals in teaching, testing, or research;
2. Periodically review the vertebrate animal care and use program and inspect the sites where animals are housed and used in research; and
3. Investigate concerns regarding the use and treatment of vertebrate animals in teaching, testing, and research at the institution.

4.3 IACUC REVIEW OF PROPOSED ACTIVITIES AND PROJECTS

As stated above, the IACUC conducts activities to ensure that animals are properly used and cared for. Before any project can begin using vertebrate animals, the

IACUC must review the proposed work to ensure concordance with regulations, policies, and sound practices. Typically, the IACUC requires the researcher to submit a complete written description of the proposed work in the form of a protocol. Certain information should be included in the protocol so that the IACUC can conduct a complete evaluation, including:

1. **Justification for animal use**. The researcher should explain why animals are needed for the work. As part of this, the potential benefit that would result, in terms of advancement of knowledge, science, or biomedicine should be explained. The IACUC should weigh this explanation against the number of animals to be used and the procedures to be conducted as a way to determine if there exists sufficient reason to use animals.

2. **Animal number justification**. The proposed number of animals to be used should be stated and clearly explained. In some cases, statistical analyses may be offered as a way to demonstrate that a certain number of animals are needed to reach a valid conclusion. In other cases, the number might be justified by the number needed as a result of the amount of tissue that needs to be collected for experiments, or by the number of animals needed to teach a specified number of students as part of a classroom learning experience. Because of the importance of selecting the best experimental design for the research while using the minimum number of animals, a separate section on sample size estimation is included at the end of this chapter.

3. **Alternatives to animal use**. When the proposed work might result in significant pain or distress to the animals, the researcher must make sure that there are no alternatives to the proposed use of animals. The consideration of alternative methods is a concept referred to as the **3Rs** and reflects the idea that in some cases the use of vertebrate animals might be improved or avoided altogether (Russell and Burch, 1959). For example, it may be possible to **replace** the use of vertebrate animals by means of other systems, such as computational models or tissue culture. Further, in some cases it is possible to **reduce** the use of animals via careful experimental design that might reduce the number of experimental groups or reduce the number of animals in each group that is to be evaluated. Finally, it may be possible to **refine** the use of animals; that is, to reduce the pain and distress that an animal might experience. Refinement can take a variety of forms, including the appropriate use of analgesics following invasive procedures, the use of methods or techniques that involve less pain or distress, or providing an enriched environment that meets the animals' needs. In some cases, alternatives may not be available or cannot be used due to interference with the scientific requirements of the work. Usually, the researcher will conduct a literature search to determine if appropriate alternatives exist for the particular project being proposed.

4. **Description of the proposed work**. A clear description of how the animals will be used in the project must be included as part of the protocol. This includes descriptions and methods to be used for any samples to be taken, compounds to

be administered, restraint of animals, surgical procedures (including steps taken to ensure aseptic technique), anesthetics or analgesics to be used, the method of euthanasia, and any other manipulations to the animal or its environment that are performed as part of the research or teaching activity. In addition, the use of any hazardous materials should be described, including steps that will be taken to protect personnel from the associated hazard. If the type of animal to be used or the procedures to be performed are likely to result in illness or debilitation of the animal, the protocol should include how the animal will be evaluated and at what criteria will be used to determine that the animal will be euthanized.

As described above, the method of euthanasia should be described, including the point at which it will occur. Unless justified for scientific reasons, the method of euthanasia must conform to the guidelines of the American Veterinary Medical Association (2013).

5. **Personnel**. Within the protocol, it must be stated who will conduct the procedures involving animals. In particular, the IACUC must determine that personnel are properly trained to handle the animals safely and responsibly. In this regard, it is common for participants in projects using animals to complete required training delivered by a variety of means, including online, classroom, and wet lab environments. It is absolutely essential that people handling animals are trained so that the animals are handled in a way that protects both personnel and animals. For example, when surgery is to be performed as part of the project, personnel should be qualified with respect to anesthesia, aseptic technique, the specific surgical manipulation being performed, and postoperative care of the animals, including management of postoperative pain. For all projects requiring euthanasia of the animal, it is essential that personnel be trained on proper methods.

Once a researcher has written the protocol, it is submitted to the IACUC for review. The IACUC must review the protocol and then decide if the use of animals as proposed is allowed to proceed. In this regard, the protocol may be approved or approval withheld (rejected). In many cases, the IACUC may have questions or determine that more information is needed, thus requiring communication with the researcher. Only after approval by the IACUC may the project begin. If any changes are made to the work described in the approved protocol, the researcher must first have the IACUC review and approve any such modifications. Importantly, the decision of the IACUC may not be changed by any other authority, including the IO.

Once the protocol has been approved and the research has commenced, it is not uncommon for the IACUC to conduct **postapproval monitoring**. This process involves assessment of the animal work to ensure that the work being conducted does not differ from that described in the approved protocol. In many cases, this is done via an IACUC representative visiting the laboratory to review records, discuss progress with personnel, and to visually observe procedures being conducted on animals. Postapproval monitoring not only provides assurance to the IACUC that approved procedures are being followed, but it also allows an opportunity for laboratory personnel to consult with someone if questions or unexpected difficulties arise. In the rare circumstance where egregious noncompliance with policies or regulations is identified, the IACUC may suspend the work involving animals.

4.4 PERIODIC REVIEW OF THE ANIMAL CARE AND USE PROGRAM AND FACILITIES

The IACUC has the responsibility to ensure that animals are being cared for appropriately and that the procedures involving animals are being performed properly and as stated in the protocol. In this regard, the IACUC is charged with conducting a **semiannual review of programs** and **inspection of facilities** related to animal care and use.

The review of programs related to animal care and use is meant to be a comprehensive consideration of related processes and procedures. The National Institute of Health (NIH) Office of Laboratory Animal Welfare (OLAW) has developed a list of items for review that IACUCs may choose to use as a tool (Office of Laboratory Animal Welfare, 2014). As part of the semiannual review, the IACUC should discuss all areas related to animal care and use, such as veterinary care, animal husbandry and environment, training of personnel, and occupational health and safety. In addition, the IACUC should consider IACUC processes and format. In the end, the semiannual program review is meant to ensure that the animal care and use program is operating within regulatory requirements and other guidelines, follows best practices within the field, and meets the needs of the animals, the personnel who care for them, and the researchers.

As part of the semiannual review, the places where animals are housed and used are typically visited and inspected by the IACUC. During these inspections, the IACUC evaluates things such as the health and living conditions of the animals, environmental conditions, the rigor of sanitation and disinfection, personnel safety, the quality and condition of feed and bedding provided to animals, and handling of hazardous agents. In addition, the areas where animals are used, such as procedure rooms or laboratories, are also inspected. At these sites, the IACUC will again review cleanliness, environmental conditions, and equipment used for animal-based procedures.

The findings of the semiannual program review and facility inspection are benchmarked against regulatory requirements and institutional policies designed to ensure consistent and optimal animal care and use. The IACUC decides if each facet of the program and facilities is acceptable; or if there is a minor or significant deficiency. A minor deficiency would be one that does not pose a significant and direct, sometimes immediate, threat to the well-being of the animals or safety of personnel. In contrast, a significant deficiency is one that may pose substantial risk to animals, personnel, or in some cases the integrity of the research.

Following the semiannual review and inspection, the IACUC produces a report that describes all noted deficiencies, classified as either minor or significant. In addition, the list includes a plan to correct each deficiency and a timetable for completion of corrections. Significant deficiencies typically need to be corrected quickly, and in some cases the activity causing the deficiency may need to be stopped until correction. The report may also note resources needed to meet anticipated needs of the animal care and use program. Once the IACUC has signed the report, it is provided to the IO for consideration and, if needed, action such as allocation of funds or other resources needed to maintain the animal care and use program at a high standard.

4.5 INVESTIGATION OF CONCERNS REGARDING THE USE AND TREATMENT OF VERTEBRATE ANIMALS

As a way to ensure that mishandling of animals is minimized, the IACUC is charged with investigating concerns involving the care and use of research animals. Such concerns might be brought to the IACUC's attention from many sources, such as someone working in a laboratory, an individual within the animal housing facility, or even a member of the public. In all cases, the IACUC must determine if there is cause for further action or sanctions, such as suspension of a protocol and cessation of animal research activities, additional training for personnel, revised procedures or processes for work involving animals, or exclusion of specific individuals from work with animals.

It is important that the IACUC has a clear and evident procedure for reporting concerns. Preferably, the procedure should be available electronically, as well as being posted within the animal facility. In addition, the introductory training for personnel should include instructions on how to report concerns to the IACUC. Methods for reporting typically include written, email, or phone. It is essential that individuals reporting concerns be protected from reprisal as whistle-blowers and that reports may be made anonymously by choice. It is common that reports may be submitted to the IACUC office, the IACUC Chair, the veterinarian, or any other member of the IACUC.

Once a report is received, the IACUC Chair will generally decide on an approach to evaluate the expressed concern. The investigation may involve formation of a subcommittee of the IACUC or in some instances the Chair or other person designated by the Chair might handle the investigation. Generally, input is sought from the person(s) identified in the report as having responsibility for the expressed concern. The IACUC will then review the concern and collect information and facts to determine if specific sanctions are merited.

4.6 SAMPLE SIZE ESTIMATION IN LABORATORY ANIMAL EXPERIMENTS

Experiments with laboratory animals are confronted with the potentially contradictory challenges of (1) minimizing the number of animals used, and possibly sacrificed, on ethical grounds and also for reasons of cost and effort, and (2) using sufficient animals such that results are statistically defensible and thus animal wastage on inconclusive experiments is minimized. Prior to undertaking any study or experiment, scientists should ask themselves four fundamental questions about their research intentions:

1. "What do I want to measure and does it involve manipulating a factor(s)?" This question addresses the **project objectives** and determines if the project is a controlled experiment (i.e., with control and treated groups) or an observational study (i.e., no attempted manipulation of groups).
2. "How do I set up my study or experiment?" This relates to the **experimental design** and entails determining the treatment levels (if applicable), number of treatment groups, and other experimental details such as housing and care in the case of laboratory animals
3. "How should I sample in space and time?" This is a **sampling design** question that covers issues of when and where the samples are taken, including the time point(s) at which the animals are sampled and whether repeated measurements are taken from the same animal.
4. "How many samples should I take?" Stated another way, "how many animals should I use?" This is most fundamentally a **statistical question** but the decision will be influenced by the ethical and cost issues mentioned previously.

Regardless of how well-thought-out questions 1—3 may be, if the experimenter does not know how many samples to take, or animals to allocate to the experiment, the remainder of the project may be meaningless. Three outcomes may result from a decision of how many animals to allocate to an experiment: (1) too few animals are allocated to the treatment groups, resulting in the inability to detect a treatment effect even if one exists (this is referred to as Type II Error in statistics); (2) too many animals are allocated to the experiment, when fewer animals would have resulted in the same interpretation (i.e., oversampling); and (3) the appropriate number of animals is devoted to the treatment groups. Cleary, two out of the three outcomes [(1) and (2)] are bad decisions based on animal ethics, cost, and effort, and experimenters should strive to estimate the most suitable number of animals to use a priori in a study [outcome (3)].

As in all scientific studies, however, the challenge in animal experimentation is to determine the "best" number of animals to allocate to the study or experiment. This question is neither simple nor straightforward, but certain statistical principles can help guide our choices. In general, the larger the number of experimental units (i.e., animals) dedicated to the treatment groups, the greater the chance of finding a statistically significant effect if such an effect truly exists. This is often called the "power" of a test or experiment. Sometimes, a spurious significant effect is revealed in an experiment that happens by chance alone and not due to the experimental treatment (known as Type I Error in statistics). However, in general, adding experimental units gains statistical power, although in an exponentially declining fashion as sample size increases. For example, increasing the sample size from 3 to 4 will gain substantial statistical power, whereas going from 98 to 99 will gain relatively little power. This "law of diminishing returns" is illustrated in Fig. 4.1.

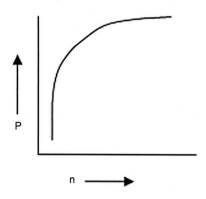

FIGURE 4.1

An increase in the sample size n will gain substantial statistical power (P) at low sample sizes (n), but a similar increase at high n will gain the experimenter little statistical power.

Fortunately, a number of approaches can help guide our selection of an appropriate sample size for a study or experiment. It is not our intention in this short treatment to delve into mathematical descriptions of sample size estimation, although a large body of information exists on this topic in the primary literature (e.g., Dell et al., 2002; Charan and Kantharia, 2013) and in textbooks (e.g., Julious, 2010; Mathews, 2010; Ryan, 2013). Rather, we will present some options and strategies that point students and new researchers in suitable directions for determining sample sizes of animals that maximize power while limiting animal wastage.

In most experiments, a control (or untreated) group of animals will be compared with a treated group of animals, or possibly several groups of animals receiving different treatments (e.g., different dosages of an experimental drug). In general, it is optimal to keep the experiment "balanced," that is to allocate the same (or similar) number of animals to each of the experimental groups including the control. The absolute minimum of animals per group will be two in order estimate experimental error, but power will be very low and larger sample sizes are typically needed to reduce Type II Error to acceptable levels. In addition, living organisms may be lost over the course of an experiment, which would further reduce sample size. To estimate requisite sample sizes, several general approaches may be used:

1. Explore the **published literature** on similar experiments or animal species to gain insight on the approximate sample size per group needed to obtain significant results (assuming a real treatment effect exists). Such a literature study may "frame" the sample size for your study but be aware that extremely significant effects (e.g., p-value less than 0.001) may suggest a larger than necessary sample size while barely significant effects (e.g., p-value about 0.05) may indicate the lower end of the recommended sample size.

2. Examine the results of **previous studies** conducted in your laboratory or animal facility with similar conditions and animal species. These studies should provide some information on the variation (i.e., "error") between animals in their response to an array of experimental treatments and conditions. High variation will suggest that larger sample sizes will be needed whereas low variation among individuals will suggest that smaller sample sizes may be needed to detect significant differences.

3. Conduct a "pilot study" with the animals of interest to better target the needed sample size for the main experiment you have designed. For example, if you plan to use several treatments in your principal experiment, such as developing a dose—response curve to a particular agent, you may wish to first compare a control group with one treated group in your pilot study using sample sizes deemed reasonable. These results will inform your later selection of sample sizes in the larger experiment to minimize animal wastage and effort and optimize experimental power.

4. Use a family of statistical approaches called "sample size statistics." These statistical algorithms typically rely on some previous information about the experimental group [such as obtained by (1—3) above] or at least "educated guesses" about population parameters. Sample size statistics can be quite informative about the sample size necessary for (1) attaining a certain power level in a two-group or multigroup experiment, or (2) revealing a particular quantitative difference between group means, as specified by the researcher, to be significant.

Most statistical textbooks contain a treatment of sample size statistics (e.g., Zar, 2010; Rosner, 2011; Sokal and Rohlf, 2011) that presents basic approaches for sampling a single population, multiple populations, or groups within a controlled experiment. In addition, various commercial or freeware software packages (e.g., "R"; http://www.r-project.org/) exist that will calculate requisite sample sizes for different scenarios. An internet search for "sample size statistics" will reveal many sites that present formulae for calculating sample sizes given the input of certain attributes of the population(s) or experimental groups. Care should be taken, however, to use reputable web sites, examine and understand the algorithms you use, and cross-compare results from different sites to increase confidence in the recommended sample size.

In summary, the selection of sample size is crucial to good experimentation. You will need to decide on the sample size for your study or experiment prior to starting the experiment, and the old adage that "more is better" does not necessarily apply to animal studies. Indeed, in animal research, it is particularly important to make informed selections that minimize animal use. Many sources of information are available for informing these decisions, including the published literature and sample size statistics that can optimize the power of an experiment while using the minimum number of animals possible.

REFERENCES

American Veterinary Medical Assocation, 2013. AVMA Guidelines for the Euthanasia of Animals: 2013 Edition. https://www.avma.org/KB/Policies/Documents/euthanasia.pdf.

Charan, J., Kantharia, N.D., 2013. How to calculate sample size in animal studies? J. Pharmacol. Pharmacother. 4, 303–306.

Dell, R.B., Holleran, S., Ramakrishnan, R., 2002. Sample size determination. Inst. Lab. Anim. Res. J. 43, 207–213.

Julious, S.A., 2010. Sample Sizes for Clinical Trials. Chapman & Hall/CRC, Boca Raton.

Mathews, P., 2010. Sample Size Calculations: Practical Methods for Engineers and Scientists. Mathews Malnar and Bailey, Fairport Harbor.

Office of Laboratory Animal Welfare, 2014. Semiannual Program Review and Facility Inspection Checklist. http://oacu.od.nih.gov/UsefulResources/FIPR_Checklist.pdf.

Rosner, B., 2011. Fundamentals of Biostatistics, seventh ed. Brooks/Cole, Boston.

Russell, W.M.S., Burch, R.L., 1959. The Principles of Humane Experimental Technique. Methuen, London.

Ryan, T.P., 2013. Sample Size Determination and Power. John Wiley & Sons, Hoboken.

Silverman, J., Suckow, M.A., Murthy, S., 2014. The IACUC Handbook, third ed. CRC Press, Boca Raton.

Sokal, R.R., Rohlf, F.J., 2011. Biometry, fourth ed. W. H. Freeman, New York.

United States Department of Agriculture, 2013. Animal Welfare Regulations. 9 CFR Part 2, §2.31 (b). U. S. Government Printing Office, Washington, D.C.

United States Public Health Service, 2015. Public Health Service Policy on Humane Care and Use of Animals, NIH Publication No. 15-8013. National Institutes of Health, Bethesda.

Zar, J.H., 2010. Biostatistical Analysis, fifth ed. Prentice Hall, Upper Saddle River.

Experimental Variables

5

K.L. Stewart
University of Notre Dame, Notre Dame, IN, United States

CHAPTER OUTLINE

5.1 THE RESEARCH ENVIRONMENT

Strict management and monitoring of all environmental parameters of a vivarium are necessary to minimize extraneous variables. Humidity, temperature, lighting, and air flow are controlled variables that must be consistently maintained with little to no variance. The facility design is the initial consideration.

5.1.1 FACILITY DESIGNS

The primary goal of the facility design is to minimize variables and to satisfy the needs of the animals and research. There are four primary types of designs; conventional, clean/dirty, barrier, and containment. The type of facility required is determined by the nature of research to be performed (Sirois, 2005).

Conventional facility design: The standard conventional facility contains a series of animal rooms, support areas, and laboratories that have single-entry doors and central corridors. Because the corridors are used for both dirty and clean caging and animals, many measures must be taken to reduce the risk of contamination between the individual areas. It is within this type of facility that it is the most difficult to manage controlled and extraneous variables.

Double-corridor facility design: Within this type of facility, also known as a clean/dirty facility, there are two doors in every room, one that opens into the clean corridor and one to the dirty corridor. All traffic in this facility is unidirectional with only clean caging and supplies entering via the clean corridor and all dirty caging and supplies exiting via the dirty corridors. Even the personnel must enter only through the clean hall and exit through the dirty hallway. In this type of facility, the chance of cross-contamination is greatly reduced.

Barrier facility: Similar in design to the double-corridor facility, the barrier facility also utilizes unidirectional traffic through a double-corridor design. However, there are many extra steps taken to ensure the health of the animals housed within. All intake air entering the heating, cooling, and ventilation system (HVAC) is passed through high energy particulate air (HEPA) filters which inhibit the passage of particles as small as 3 μm. All of the barrier areas are positive in air pressure to the surrounding areas. There are protocols in place regarding the staff; they are required to shower prior to entry and to wear autoclaved uniforms. All items, including feed, bedding, and caging, entering the area are sterilized prior to entry by irradiation, steam sterilization, or ultra-violet light exposure. Entrances often have air showers that in theory remove particles from clothing for those entering the facility. Access is limited to essential personnel only. This facility prevents the introduction of unwanted pathogens into the animal rooms.

Containment facility: The construction and practices of the containment facility is very similar yet opposite to a barrier facility. The air leaving the facility is passed through the HEPA filters. Staff members shower as they leave the facility and their clothing which is either reusable or disposable, is autoclaved. All items; trash, dirty caging, and carcasses are autoclaved prior to disposal. This type of facility is used to contain known pathogens that are being utilized in the research experiments.

5.1.2 ENVIRONMENTAL PARAMETERS

The experimental animals' environmental parameters within the research facility are placed into two categories; the microenvironment and the macroenvironment. The microenvironment represents the conditions within the animal's primary enclosure while the macroenvironment represents the conditions in the secondary area, the animal room. All of the parameters must be set and maintained in accordance with the needs of the animals and within the guidelines set forth by the *Guide for the Care and Use of Laboratory Animals (the Guide)* (Institute for the Laboratory Animal Research, 2011) and the Animal Welfare Act Regulations (AWAR) (United States Department of Agriculture, 2013). Preferably, all parameters are monitored by a system that not only records the time points but also has call-out options that alert facility personnel of fluctuations in the parameters.

5.1.2.1 Macroenvironment

Macroenvironment consists of the following: the construction materials used throughout the facility; air quality which includes ambient temperature, humidity, ventilation rates and pressure differentials; noise and vibration within the animal rooms and surrounding areas; sanitation practices; and the animal care staff.

Animal room construction: All of the building materials must be easily cleaned and durable. There should not be any seams in the floors or on the wall surfaces. Animal room doors should have a viewing window to allow for observations of the room prior to entry. Doors should open into the room to contain any escapee animals. In nonhuman primate areas, special precautions are taken to prevent animals that have escaped from their primary cage from getting into duct work or electrical boxes. For example; lights are recessed with protective covers placed over the entire unit, exhaust grills are bolted with security fasteners rather than standard screws to prevent the monkeys, upon escape, from getting into the lighting components. Also, cages are attached to wall-mounted bars to prevent the cage from moving.

Air quality: The air within the animal room must be maintained within the comfort zone of each species. This includes the proper temperature, humidity, air exchange rates, and pressure differentials. Animals of different age and health status may need different parameters, especially for the ambient temperature. The animal rooms should be set at the animals' thermoneutral zone, the temperature range that allows the animals to maintain homeostasis without expending energy. The optimal temperature ranges are 68-72°F for most species; however, some species may require higher or lower temperatures dependent on their physiological status. The optimal humidity range for most species falls in the 40–60% range. Excessively low humidity can result in ringtail in mice and rats, a condition where rings of contracted skin form around the tail to the point of tissue constriction and necrosis. If the humidity is too high, it can result in an increase in ammonia levels which can have negative effects on animal health. Although in many facilities there will be zone settings for the humidity levels, each animal room should have its own thermometer to allow for species-specific requirements. See Table 5.1 for the species-specific requirements for temperature and humidity.

The air exchange rate is specified in the Guide at a minimum of 10 air room changes per hour. As air is exchanged, there is a decrease in the number of microbes in the air while also decreasing the odors in the room. However, too many air changes per hour could create drafts that might affect the animals' ability to thermoregulate.

The pressure differential requirement is based on the type of facility and the purpose of the research within the room. For areas that house breeding colonies, specific-pathogen-free (SPF) animals or animals that are immunocompromised, the air differential is set so that the animal room has positive pressure to the surrounding areas. The positive pressure forces air out of the door as it is opened, preventing potentially contaminated air from coming into the room. Conversely, in areas that house known pathogens, the air differential is negative to the surrounding areas, thus preventing air from exiting the room and allowing contaminated air to escape into noncontaminated areas.

Table 5.1 Recommended Ambient Room Temperature for Commonly Used Laboratory Animals

Species	Temperature Range (°F)	Temperature Range (°C)	Relative Humidity (%)
Rodents	64–79	18–26	40–60
Rabbits	61–72	16–22	30–70
Dogs, cats, primates	64–84	18–29	30–70
Chickens—adult	61–81	16–27	30–60
Chicks—1 to 10 days	95–100	35–38	40–60
Zebrafish (to maintain the temperature in the water)	80–85	27–29	N/A

Adapted from Institute for the Laboratory Animal Research., 2011. Guide for the Care and Use of Laboratory Animals, eighth ed. National Academies Press, Washington (DC).

Lighting requirements: Generally animal rooms are built with no external windows thus in most circumstances illumination is often from internal lighting. The lights should be on a scheduled cycle, normally 12:12 h on and off. However, there are some circumstances that require a different schedule such as for some species of breeding animals or for experimental protocols.

The intensity of the light is crucial. Each room should have at least two levels of lighting, one to allow adequate lighting for cleaning, 70–100 foot-candles, and one that is comfortable for the animals, 30–50 foot-candles. Lighting must be evenly distributed throughout the room. All lighting should be monitored to assure proper and consistent functioning. Albino animals are prone to retinal degeneration when exposed to lighting levels over 50 foot-candles for prolonged periods of time.

The use of an astrological clock with low-level lights automatically controlled by means of a rheostat allows for the simulation of dusk and dawn and can be set to any location in the world by inputting the longitude and latitude. Animals that are wild caught or that have a biological need to experience dawn and dusk benefit greatly from this type of lighting system.

Noise and vibrations: Animals are very sensitive to noise, including frequencies well beyond the scope of the human ear. The equipment used throughout the animal facility can emit high-frequency sounds and vibrations that can interfere with breeding of mice, rats, and even zebrafish. Mouse and rat breeding colonies should be housed in rooms that are positioned away from cage-wash areas, elevators, common use rooms such as procedure rooms, and away from other species that are loud such as dogs and primates (Reynolds et al., 2010; Norton et al., 2011).

Sanitation practices: A critical aspect of animal health and well-being is maintaining a sanitary and meticulous environment through the use of proper cleaning procedures and safe and effective cleaning chemicals. The animal facility has the potential of harboring a myriad of microorganisms, many that could be harmful to the animals and detrimental to the integrity of the research data collected. Because there are a variety of soil types, namely oils, urine scale, hair, dander, and fecal matter, an assortment of cleaning agents are employed. The cleaning chemicals that are used in the animal rooms must be carefully selected as they can be toxic to many species. For example, phenolic agents that are found in many cleaning products are toxic to cats. Thus, all chemicals brought into the vivarium, including those used by the custodial staff, must be approved by the facility management (Hidell, 2003; Ingraham et al., 2013).

Cleaning procedures include the removal of excessive dirt, excrement, and debris from the surface to be disinfected. The surface is then treated with a cleaning agent, an acid or alkaline solution that is soluble in water. Acidic solutions remove urine scale while the alkaline detergents remove proteinaceous soils from the surface to be disinfected. Chemical disinfectants have germicidal properties that disrupt microorganism cells through membrane damage thus rendering the microbe harmless.

The frequency of cleaning of the animal rooms and caging implements is determined by a variety of parameters; species housed, cage densities, bedding types, environmental conditions, and types of enclosures. The Guide has set forth suggested schedules for the cleaning of enclosures and caging implements such as water bottles and feeders; enclosures and accessories. All should be sanitized at least once every two weeks; however, solid-bottom caging, sipper tubes, and bottles are usually in need of cleaning once weekly. It is the responsibility of the management to establish standard operating procedures for all scenarios within the animal facility.

Animal caretakers: It has been demonstrated that the simple act of handling an animal, such as moving it from a dirty cage to a clean cage, can activate the hypothalamus−pituitary−adrenal axis (HPA axis), causing an increase in corticoid steroid levels. Increase corticoid steroid levels are indicative of stress. Continual spikes in corticoid steroid levels can impact research studies (Balcombre et al., 2004). Thus the importance of a well-trained animal care staff cannot be denied.

Studies have demonstrated that animals are able to discriminate between individuals and will associate the individuals with events that occur in their presence (Davis, 2002). Because of this ability, it is common for strong bonds to form between the research animals and the animal care staff. Similar to classic Pavlovian conditioning, animals learn to predict the events that will occur during the presence of individuals and can react appropriately to the stimuli caused by that individual. Thus the stress levels induced by routine procedures are minimized by having consistency in the animal care staff and procedures (Bayne, 2002).

5.1.2.2 Microenvironment

Microenvironment includes the following: caging type, food, water, bedding type, pheromones, housing densities, and pathogens present in the animals.

Caging type: There are a variety of caging materials used for the primary enclosures for animals, mainly stainless steel, aluminum, and plastics. The type of material to use is determined by the species to be housed. Caging made from stainless steel is durable, smooth, and easily cleaned. However, it is also very costly, heavy, and draws heat away from the animals. Aluminum is a less expensive alternative to stainless steel, but is less durable, and is easily damaged by larger animals. The use of highly refined plastics that can withstand chemical disinfectants and steam sterilization has become very popular for most species, especially rodents.

Individually ventilated caging is commonly used for housing rodents, mainly mice. These cages are seated onto a rack that has air introduced and diffused at the cage level at a low velocity which keeps the bedding dryer longer, and prevents large amounts of ammonia from being created (Ferrecchia et al., 2014). The air introduced in the rack is HEPA filtered. These racks are designed to house a large number of animals in a very small space, increasing the capacity of the animal rooms.

Caging has been designed to meet specific research needs. Metabolic cages are structured to collect urine and feces. All feed and water implements are located such that there is no contamination in the collection vessels. Caging can also be designed to allow group housing or socialization of animals. Cages are also made specifically to transport animals to laboratories or diagnostic facilities.

Caging densities: The minimum space requirements for each species are mandated by the Guide and the Animal Welfare Act Regulations. These requirements are based on the behavioral and biological needs of each species. The caging must allow animals to perform normal postural adjustments and movements. Physiological factors such as recovery from illness or surgical procedures, reproduction, or growth can be affected by the caging density. These requirements must be reassessed throughout the duration of the research project as the needs of the animals may change as the project proceeds and animals grow.

As animals are received for an experiment, they should be initially housed as they will be throughout the experimental process. Animals will form bonds which if interrupted can cause stress to the animals. Male mice, because of their inherent aggression, can be a challenge to group house. Much research has been done to modify the aggression in the male mice. Housing density has been shown to be a contributing factor. The ideal group size of three male mice per cage has been established for one strain of mice (Van Loo, 2001a).

Food and feeding devices: The primary animal diet must be based on the biological and physiological needs of each species. Feeders must be designed to minimize waste, allow for easy access, and minimize the potential for contamination by urine and feces. Feed bowls on the cage floor are ideal for some species whereas hanging feeders suspended in the cages are best for others.

The food must be palatable and available on a regular basis. The diet must meet the nutritional requirements of the animals. Also the shape and texture of the food must be acceptable. Rodents have open-rooted incisors that grow continuously. They require large, hard, and dense food pellets to maintain the correct incisors length.

There are two types of feed; open formula and closed formula. Within the closed formula there are fixed formulas and constant formulas. An **open-formula** diet's ingredients and formulation are publicly available allowing researchers to easily duplicate the diet. All changes in the diet composition are also disclosed to the public. A **closed-formula** diet is manufactured under a company trade name thus making the formulation proprietary. Because the formulas of closed diets are not disclosed, changes to the diets can be made without notification to the customers. This includes the use of lesser-quality protein sources, depending on the market costs of the ingredients. The **constant formula** is a diet that is formulated with consistent concentrations of nutrients with the group of the ingredients remaining unchanged. However, the combination of the percentage of each ingredient within a group can vary. As an example, a manufacturer may guarantee a diet with 12% protein supplied by a group of ingredients that consist of corn, oats, fish meal, wheat, and soy. However the exact combination and percent of each ingredient can vary according to the supplies on hand at the time of milling. The **fixed formula**, however, always maintains the percentages of the ingredients in all nutrition groups. The constant and fixed diets can be either open- or closed-formula diets (Barnard et al., 2009).

Animals that are immunocompromised require sterile food. Commercial suppliers offer feed that is either irradiated prior to shipping or food that can be steam sterilized at the animal facility. Food that is to be steam sterilized must be formulated such that the vitamin content is not compromised. The sterilization process can also cause changes in the texture and palatability. Although more expensive, the advantages of having feed irradiated typically offset the additional costs and is used by most institutions. For FDA studies, certified diets are required. Although identical to noncertified diets, each batch is tested and certified to be free of all contaminants prior to shipping.

Water: Animals must have access to a constant supply of fresh potable water. This can be delivered via water bottles or through an automatic watering system. Advantages to using water bottles include the ability to add medications or test agents into the water and to measure water consumption. Although the cost of the bottles is minimal, the use of them is very labor intensive. The automatic watering system is initially expensive but requires minimal labor once installed. The animals have continual access to fresh water. However, there are system failures such as flooded caging from a leaking sipper tube or flooding in the animal rooms due to a broken pipe. Experimental agents cannot be added to the water nor can the water consumption be quantified.

Bedding: The bedding materials not only affect the animal's well-being by providing comfort and thermal regulation, but can also influence experimental results. Important factors to consider when selecting bedding are availability, comfort for the animals, absorbency, and disposability. In addition, bedding should be nontoxic to the animals and nonnutritive. The common types used for laboratory animals are: ground corn cobs, shredded or chipped wood, and paper products that are made specifically for use in rodent caging. The choice of which one to use is commonly determined by the animal facilities management. It is the responsibility of the Principle Investigator (PI) to know which beddings are used and how they will affect the results of their studies. Some studies may require flexibility with respect to the bedding. For example, when a study has a nutritional component, it is counterproductive to house the animal on corn cob bedding as it often contains small pieces of corn. The animal's consumption of these corn pieces can alter the test results (Ago et al., 2002; Kawakami et al., 2007).

A key factor in the choice of bedding is absorbency as it affects the frequency of cage changing. Combinations of bedding, absorbent bedding like corn cob and more comfortable paper chips are often used in the **individually ventilated cages** (IVCs). Nesting materials such as shredded kraft paper and compressed cotton fibers are often added to allow the animals to thermoregulate, to build nests for litters, and to opt to be out of the light and the air flow.

Soiled bedding must be replaced with new bedding as needed to keep the cage environment dry and free of ammonia. Typical IVCs are changed every 10−14 days while static cages are changed every 3−4 days. The frequency of the bedding changes is dictated by several factors; the species, the number of animals, the type of bedding, and environmental factors in the animal room such as temperature and humidity. The experimental conditions such as animals that have recently had surgery or breeding colonies may impact the frequency of bedding changes (Domer et al., 2012). Bedding changes can also contribute to aggression in cages of male mice. Placing the nesting materials from the dirty cage into the clean cage has shown to lessen the territorial fighting in the male mice (Van Loo, 2001b).

Pheromones: Pheromones are chemosignals that produce olfactory cues for intraspecies communication. Usually released in urine, these signals are important in behavior and reproduction. In mice, males produce pheromones that cause the Whitten effect, which stimulates the synchronization of estrus in females. Another hormonally induced phenomenon, the Bruce effect, is when the introduction of an unfamiliar male into a cage with a pregnant female causes her to terminate the pregnancy. Other pheromones can induce aggression (Novotny et al., 1995; Morè, 2008). The use of IVCs greatly diminishes the incidence as each cage is individually supplied by air and minimal cross-contamination of air occurs.

Pathogens: Pathogens in the form of bacteria and viruses must be controlled in the animal rooms. Animals of known health status should be used for biomedical experiments. Pathogens inadvertently introduced during an experiment can alter the results and skew the data. Care must be taken to avoid accidental infection of experimental animals with pathogens. See Section 5.3 of this chapter.

5.2 ENVIRONMENTAL ENRICHMENT

As the concept of environmental enrichment became popular for captive animals, the Association of Zoos and Aquariums (AZA) set forth the following definition of enrichment: "Environmental enrichment is a process for improving or enhancing animal environments and care within the context of the inhabitants' biology and natural history. It is a dynamic process in which changes to structures and husbandry practices are made with the goal of increasing behavioral choices available to animals and drawing out their species-appropriate behaviors and abilities, thus enhancing animal welfare." (Hare, 2002) This section will discuss how the goal of environmental enrichment is met in the laboratory animal setting.

In standard animal caging, laboratory animals are unable to express their highly motivated behaviors, which often lead to the development of stereotypic behaviors such as wire bar lid gnawing, pacing, or excessive grooming (Wurbel. 1998; Hart et al, 2009). It has been demonstrated that animals raised in an enriched environment have improved learning and problem-solving ability (Renner et al., 1987) and better indices of emotionality and environmental adaptation (Larsson et al., 2002). Another study demonstrated that there is striatal dysfunction, which has been implicated in the development of obsessive disorders in humans, in the animals that are unable to act upon their natural preferences. Animals that have dysfunctional stereotypies are not good models for function, pharmacologic, and genetic studies (Garner and Mason. 2002). Standardizing species-specific environmental enrichment enhances the animals' overall well-being and reduces the animal variability that stereotypic behaviors can introduce into a study.

5.2.1 ENRICHMENT PROGRAM

Enrichment program development is the responsibility of the animal care management staff and the attending veterinarian with the consultation of the researchers and ideally with a behaviorist, if one is available. It is helpful to develop enrichment plans for each species housed within the facility. Enrichment committees can be established to consider enrichment options, establish the feasibility of each option, design or build the prototypes of enrichment devices if needed, and then to establish the implementation and evaluation of each strategy utilized. Members of the animal care staff and veterinary team should be included in each stage of the program. Upon implementation of new enrichment strategies, the animal care staff is able to make observations built on the baseline behavior and the animals' reaction to novel objects placed within their enclosure. An unfamiliar observer can skew the results as the animals could be reacting to the person rather than the new enrichment. Notations should be made on the animals' behavior prior to introduction of the item, within the first hours, and also after the animals have become adjusted to the presence of the item. An object that creates an immediate reaction and encourages exploration, but soon is completely ignored or one that elicits a strong fear response is not a successful enrichment strategy. Effective plans encourage the animals to utilize their natural abilities through exploration and locomotion (Stewart, 2003, 2004; Stewart and Bayne, 2004; Weed and Raber, 2005; Wolfe, 2005).

5.2.2 SOCIAL ENRICHMENT

Social enrichment is provided in two ways: social housing within a space that involves pair or group housing and which meets the social needs of the animals through direct contact with conspecifics (members of the same species); or nonsocial housing that provides visual, olfactory, and auditory cues from conspecifics housed in the same room (Fig. 5.1). For situations that require animals to be singly housed, sensory cues may serve as a substitute for direct contact. Although singly housed

FIGURE 5.1

Although individually housed, this rabbit's enriched environment has a large floor space for exploration and many novel items with which to interact. Because other animals are housed within the room, auditory, visual, and olfactory cues provide social interaction despite the inability for direct contact.

animals are not able to partake in normal species-specific physical interactions, they still may remain aware of their conspecifics and are able to react to the sensory cues provided by them. Because appropriate social interactions with conspecifics are essential to the normal growth, development, and well-being of animals, all social species should be paired or group housed unless otherwise justified by scientific or animal health reasons or because the animals are incompatible.

5.2.3 NONSOCIAL ENRICHMENT

Nonsocial enrichment strategies provide animals with structures and husbandry practices that encourage animals to exhibit their normal behavior. For example, providing opportunities for nest building, burrowing, foraging, and perching will encourage normal species behaviors and decrease pathologic maladaptive behavior or stereotypies (abnormal repetitive behaviors that serve no purpose) (Smith and Corrow, 2005). Such behaviors are believed to result from barren and unchallenging surroundings that do not allow animals to perform behaviors within their natural repertoire.

Studies have demonstrated that animals have reduced stress if they are provided with the opportunity to have control of their environment. One way to achieve this is by providing nesting materials or housing structures. Nesting materials such as EnviroDri (Shepherd Specialty Papers, Kalamazoo, Michigan) and Nestlets (Ancare, Bellmore, New York) are commercially available. Tissues, paper towels, and paper tubes are also sometimes used. Providing these materials allows rodents to satisfy their natural behavioral of burrowing and nest building (Fig. 5.2). There are many commercially available enrichment devices made of transparent plastic or cardboard for rodent caging. For example, shelters permit the animals to avoid the lighted area of the cage, escape from aggressive cage mates, and create an

FIGURE 5.2

Social enrichment of mice housed together in an enriched environment. Providing a tube will enhance exploration. The plastic hut provides an area for of shelter, while shredded nesting materials enable the mice to utilize their natural burrowing and nest-building behaviors.

area for sleep. It has been demonstrated that the use of a cardboard hut in a cage of male mice reduces aggression if the hut is transferred from the dirty cage to the clean cage upon cage changing. It is believed that the hut contains only calming pheromones that the mice emit during periods of rest (Van Loo, 2001c).

Other enrichment strategies that allow control for the animals include varying food presentation, providing perches in the enclosures, and offering toys or other items that the animals can manipulate. In their natural setting, animals spend a significant amount of time hunting for food. They must utilize their olfactory, auditory, and visual capabilities to acquire their food. Studies have shown that animals prefer to work for their food, even when food is readily available (Garner, 2005). Scattering seeds within rodent cages provides foraging opportunities that promotes both mental stimulation and locomotion. Mixing the feed for chickens with a substrate such as aspen shavings allows them to display their natural hunt and peck feeding behavior. Supplying puzzle feeders or placement of food in a variety of locations within the enclosure of nonhuman primates requires the animals to use visual cues and manual dexterity to attain the food. Providing whole fruits and vegetables or nuts in the shell can increase feeding time, increase overall feed intake, and increase dietary diversity for nonhuman primates.

Some animal species benefit from having a perch or shelf within the cage. For example, chickens need a perch that allows the natural posture of their feet while sleeping. Cats housed in a communal setting need perches or shelves at varying heights to allow them to establish the hierarchy in the colony. Many of the nonhuman primates are naturally arboreal, thus need to have resting boards placed above the ground level.

Enhancement of the enclosures of animals that require a more stimulating environment can prevent maladaptive behaviors. Utilizing toys, puzzles, or other objects that can be manipulated by the animal encourages exploration and play. There are many useful devices that are commercially available for this purpose. For example, plastic balls are rolled around and rattle-type toys are manipulated and tossed around the cage by rabbits. Nonhuman primates will show preferences for toy type and even colors. However, such implements must be carefully selected. All items must be made of nontoxic materials, be sanitizable or disposable, free of sharp edges or places where digits or limbs could become trapped, easily installed or placed in the cage, and sturdy enough to withstand use by the animals. Safety concerns for both the animals and the caretakers must also be considered. Devices should be closely examined for loose parts that could be swallowed or broken parts that could be used as a weapon by the animal, especially nonhuman primates. A balance between the cost and difficulty of installation of an enrichment device should be balanced with the benefits to the animals (Stewart and Bayne, 2004).

5.2.4 HABITUATION

Habituation of animals to procedures allows the animals to predict and acclimate to events that occur during a research project. When animals are trained through operant conditioning to present a limb for injection or blood sampling, for example, the procedures are less stressful to the animals and the animal technicians. Operant conditioning is defined as a form of learning that utilizes consequences as a means to shape behaviors. Positive reinforcement, such as treats or play time, is given to the animal when the desired behavior is exhibited. As the animal relates the treat to the behavior, whether it is extending a limb for blood sampling or standing quietly for an exam, the animal voluntarily participates in the procedure (Murphy and Lupfer, 2014).

Acclimation to a procedure is also beneficial. Allowing an animal to enter and exit a restraint device with no procedure taking place allows the animal to relax in the restraint device. Exposing the animal to a new piece of equipment without the use of it can desensitize the animal to a new sight, sound, or smell that it will encounter in future experiments. Acclimation combined with operant conditioning results in animals that are less stressed throughout experimental procedures.

5.3 ANIMAL HEALTH MONITORING PROGRAM

A key component of controlling variables within an animal study is a comprehensive veterinary medical program that maintains the health of the animals. Such a program includes policies and established protocols on the quarantine on new arrivals and isolation of sick animals, the separation of animals by species, health status, and the source of the animals. The use of animal transfer stations when manipulating mice and rats minimizes the chance of exposing animals to environmental conditions and to animals outside their enclosure. This section will define preventative measures that are taken to insure the health of the research animals.

5.3.1 ANIMAL HEALTH STATUS

SPF animals are widely used for biomedical research. The development of the SPF animals began in the 1940s with the creation of axenic (germ-free) animals at the Laboratory of Bacteriology at the University of Notre Dame (LOBUND). Colonies of hand-reared mice and rats were sustained under sterile conditions. As the mouse and rat colonies began to thrive, their value as a research tool was quickly recognized. Because of the germ-free status, these colonies were free of the common pathogens that plagued the mice and rats used in research. The removal of pathogens eliminated a fundamental variable in the animals, allowing for more uniformity of the research subjects (Pleasants, 1965).

These first colonies of germ-free animals were the start of colonies that would be maintained as SPF colonies. Although not reared under sterile conditions, SPF animals are raised under strict procedures to prevent the introduction of pathogens. The animals are defined by the pathogens that have been excluded from the colonies. Commercial breeding facilities typically provide a detailed list of the pathogens for which they test and a report on the latest testing with each shipment of animals. Careful attention to these reports by the facility management is crucial to the integrity of the health of animals housed in their facility.

5.3.2 PREVENTIVE HEALTH PROGRAM

Animal health monitoring begins with the establishment of a comprehensive vendor list that has been approved by the attending veterinarian and ordering policies. Approved vendors have established long-term disease-free colonies that can be introduced directly into the animal facility without an extensive quarantine period. Animals are often allowed a minimum of 48 h but preferably 7 days to acclimate to the facility and to recover from the stress of travel. Subclinical infections and physiological perturbations can result from the stress of travel, thus animals should be carefully monitored upon arrival. Animal orders should be placed by a person that has direct contact with the animal facility to ensure that there are adequate resources such as staffing and housing. It must also be confirmed that the animals are on an approved Institutional Animal Care and Use Committee (IACUC) protocol.

An animal health quality control program is created to maintain the health of all of the research animals housed in the vivarium. Programs are designed to detect infectious agents that commonly occur in laboratory animals. The decision of which agents to monitor for and the frequency of testing is made by the veterinary team and management. Typically, facilities test quarterly for the most common pathogens; viruses, bacteria, and parasites through serology, polymerase chain reaction (PCR), bacterial cultures, pelt exams, and necropsies followed by histopathology of tissues and organs. One method of sampling is housing animals of known pathogen status, referred to as sentinels, within the colony to be tested. The sentinels are housed within the same room and are sometimes exposed to the test animals via dirty bedding transfer. When exposed to the dirty bedding, the sentinels would, theoretically, become infected with any disease agents present. Another method used in conjunction with IVCs is to swab the air plenums of the rack and then test the swabs via PCR for the various pathogens. Further, technology has been developed that allows a media panel to be placed in the path of the exhaust air dust on the IVC rack, thus exposing it to all of the air circulating within the plenums. Because the IVCs provide cage level containment, the latter two methods are shown to be more sensitive and provide more accurate findings. The age and immune status of the sentinel animals is crucial. For a mouse colony, the sentinel should be a young immunocompetent mouse that is susceptible to the common infectious agents. Sentinels should be exposed to test animals for a minimum of 6 weeks but preferable one is 10–12 weeks to assure that they seroconvert to any pathogen present (Mähler et al., 2014).

Another possible route for introducing pathogens into a colony is through the use of cell lines. Upon any occasion that an animal is injected with cells that have been derived from another source, there is the possibility that pathogens that the donor animal was harboring will be transferred to the recipient. Thus all cell lines should be tested for common murine viruses.

Scheduling examinations, vaccinations, routine deworming, and dental prophylaxis is included in the preventive medical program for the larger mammals including; rabbits, dogs, cats, and nonhuman primates. The examinations should be performed at least annually or as needed to monitor the health of the animals.

Quarantine and isolation: Animals that are transferred from other animal facilities or collaborating laboratories should be quarantined for a minimum of 6 weeks upon receipt. Prior to release from quarantine the animals should be tested for all common murine pathogens as described in the animal health quality control program. Either animals received or sentinel animals placed with them are used for the testing. Animals that test positive for a pathogenic agent should not be introduced into the general population. Policies should be established for handling of animals determined to be infected. Isolation of sick animals is imperative until a definitive diagnosis is established. Animals that are harboring infectious diseases must be removed from the general population to prevent the spread of disease.

Separation by species: As per the regulations and guidelines, animal rooms should house only one species to prevent disease transmission between species. The separation also prevents behavioral and physiological responses in the animals that are caused by interspecies conflicts. However, there are circumstances where housing of two or more species in the same room can be justified. For example, the use of individually ventilated caging or cubicles that have unidirectional air flow and is directly vented out of the room minimizes the chances of interspecies conflict from occurring.

Preventive protective equipment is utilized by the animal care staff while working with the animals to protect themselves and the animals. At minimum, gloves and clean laboratory coats or scrubs should be required. The lab coats should remain in the animal facility and laundered regularly. Gloves should be changed between cages if there are differences in the health status of the animals. Due to the high incidence of laboratory animal allergies among research and animal care staff, masks should also be available. Some facilities will handle the animals with forceps that are disinfected between animals or cages rather than with gloved hands.

Specialized equipment has been developed to minimize the exposure of the animals to other animals and to the room conditions. As disease-free animals were developed, caging to maintain that health status had to also be designed. Originally cages were wrapped in cotton-type filter material to filter the air entering the cage. The caging progressed to a filtered bonnet top over the cage. In the 1990s, individually ventilated caging for rodents was developed. Air is introduced into the cages at a low velocity, providing a circulation of air across the bedding material.

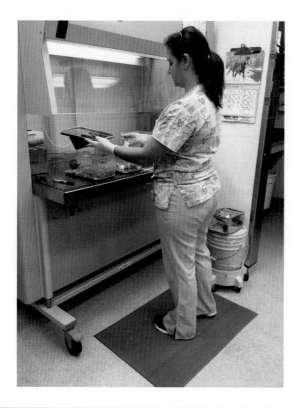

FIGURE 5.3

Changing rodent cages under an animal transfer station protects the animals from potential airborne pathogens while protecting the technician from allergens.

Consequently, the bedding remains dryer and there is a significant reduction in ammonia production. The air is HEPA filtered both on intake and exhaust. Each cage is essentially isolated from all others.

Animal transfer stations and laminar flow hoods are often utilized when a cage is opened (Fig. 5.3). The air within both the transfer station and the hood is HEPA filtered in and exhausted into the room. The steady uniform velocity body of air forces a constant positive air pressure throughout the work space, protecting the animals from the environmental air. With the opening of a singly cage within the hood, there is no opportunity for cross-contamination to occur.

5.4 SUMMARY

There exists a variety of parameters within the animal facility that must remain relatively constant to prevent unwanted or extraneous variables from being introduced into an animal experiment. The well-being of the animals is greatly influenced by environmental factors within the animal room such as humidity, temperature, lighting, and air flow. Caging design and complexity, feed, water, and even the animal care staff also impact the experimental animals. Controlling these factors prevents variables that can negatively impact the data obtained from the animals.

REFERENCES

Ago, A., Gonda, T., Takechi, M., Takeuchi, T., Kawakami, K., 2002. Preferences for paper bedding material of the laboratory mice. Exp. Anim. 51 (2), 157–161.

Balcombre, J.P., Barnard, N.D., Sandusky, C., 2004. Laboratory routines cause animal stress. J. Am. Assoc. Lab. Anim. Sci. 43 (6), 40–51.

Barnard, D.E., Lewis, S.M., Teter, B.B., Thigpen, J.E., 2009. Open- and closed- formula laboratory diets and their importance. J. Am. Assoc. Lab. Anim. Sci. 48 (6), 709–713.

Bayne, K., 2002. Development of the human-research animal bond and its impact on animal well-being. Inst. Lab. Anim. Res. J. 43 (1), 4–9.

Davis, H., 2002. Prediction and preparation: Pavlovian implications of research animals discriminating among humans. Inst. Lab. Anim. Res. J. 43 (1), 19–26.

Domer, D.A., Erickson, R.L., Petty, J.M., Bergdall, V.K., Hickman-Davis, J.M., 2012. Processing and treatment of corncob bedding affects cage-change frequency for C57BL/6 mice. J. Am. Assoc. Lab. Anim. Sci. 51 (2), 162–169.

Ferrecchia, C.E., Jensen, K., Van Andel, R., 2014. Intracage ammonia levels in static and individually ventilated cages housing C57BL/6 mice on 4 bedding substrates. J. Am. Assoc. Lab. Anim. Sci. 53 (2), 146–151.

Garner, J.P., Mason, G.W.J., 2002. Evidence for a relationship between cage stereotypies and behavioural disinhibition in laboratory rodents. Behav. Brain Res. 136, 83–92.

Garner, J.P., 2005. Stereotypies and other abnormal repetitive behaviors: potential impact on validity, reliability, and replicability of scientific outcomes. Inst. Lab. Anim. Res. J. 46 (2), 106–117.

Hare, V.J., 2002. The latest definitions. Shape Enrich. 11 (1), 1.

Hart, P.C., Berner, C.L., Dufour, B.D., Smolinski, A.N., Egan, R.J., LaPorte, J.L., Kalueff, A.V., 2009. Analysis of abnormal repetitive behaviors in experimental animal models. In: Warnik, J.E., Kauleff, A.V. (Eds.), Translational Neuroscience. Hauppauge: NY.

Hidell, T.B., Parker, A., Wilfred, A.G., 2003. Environmental monitoring: the key to effective sanitation. Lab. Anim. 32 (5), 26.

Ingraham, A., Lynch, F.E., Shapiro, K.B., 2013. Sanitation Chemicals for Laboratory Animal Science. http://www.alnmag.com/articles/2013/01/sanitation-chemicals-laboratory-animal-science.

Institute for the Laboratory Animal Research, 2011. Guide for the Care and Use of Laboratory Animals, eighth ed. National Academies Press, Washington (DC).

Kawakami, K., Shimosak, S., Tongue, M., Kobayashi, Y., Nabika, T., Nomura, M., Yamada, T., 2007. Evaluation of bedding and nesting materials for laboratory mice by preference tests. Exp. Anim. 56 (5), 363–368.

Larsson, F., Winblad, B., Mohammed, A.H., 2002. Psychological stress and environmental adaptation in enriched vs. impoverished housed rats. Pharmacol. Biochem. Behav. 73, 193–207.

Mähler, M., Berard, M., Feinstein, R., Gallagher, A., Illgen-Wilcke, B., Pritchett-Corning, K., Raspa, M., 2014. FELASA recommendations for the health monitoring of mouse, rat, hamster, guinea pig, and rabbit colonies in breeding and experimental units. Lab. Anim. 48 (3), 178–192.

Morè, L., 2008. Intra-female aggression in the mouse (*Mus musculus domesticus*) is linked to the Estrous cycle regularity but not ovulation. Aggressive Behavior. 34, 46–50.

Murphy, E.S., Lupfer, G.J., 2014. Basic principles of operant conditioning. In: The Wiley Blackwell Handbook of Operant and Classical Conditioning. West Sussex, UK.

Norton, J.N., Kinard, W.L., Reynolds, R.P., 2011. Comparative vibration levels perceived among species in a laboratory animal facility. J. Am. Assoc. Lab. Anim. Sci. 50 (3), 635–659.

Novotny, M.V., Xie, T.M., Harvey, S., Weisler, D., Jemiolo, B., Carmack, M., 1995. The stereoselectivity in mammalian chemical communication: male mouse pheromones. Experientia 51 (7), 738–743.

Pleasants, J.R., 1965. History of germfree animal research at Lobund laboratory, biology department. Proc. Indiana Acad. Sci. 75, 220–226 (Indianapolis, IN).

Renner, M.J., Rosenzweig, M.R., 1987. Enriched and Impoverished Environments. Effects on Brain and Behavior. Springer-Verlag, New York.

Reynolds, R.P., Kinard, W.L., Degraff, J.J., Leverage, N., Norton, J.N., 2010. Noise in a laboratory animal facility from human and mouse perspectives. J. Am. Assoc. Lab. Anim. Sci. 5, 592–597.

Sirois, M., 2005. The research environment. In: Laboratory Animal Medicine: Principles and Procedures.

Smith, A.L., Corrow, D.J., 2005. Modifications to husbandry and housing conditions of laboratory rodents for improved well-being. Inst. Lab. Anim. Res. J. 46 (2), 140–147.

Stewart, K.L., 2003. Environmental enrichment program development: hurdling the common obstacles. Anim. Tech. Welfare 2 (1), 9–12.

Stewart, K.L., 2004. Development of an environmental enrichment program utilizing simple strategies. Anim. Welfare Inf. Cent. Bull. 12 (1–2), 1–7.

Stewart, K.L., Bayne, K.A., 2004. Environmental enrichment in laboratory animals. In: Reuter, J.D., Suckow, M.A. (Eds.), Laboratory Animal Medicine and Management. IVIS. http://www.ivis.org/advances/Reuter/stewart/chapter1_frm.asp?LA=1.

United States Department of Agriculture, 2013. The Animal Welfare Act and Animal Welfare Regulations.

Van Loo, P., 2001a. Modulation of aggression in male mice: influence of group size and cage size. In: Male Management: Coping with Aggression Problems in Male Laboratory Mice. Surrey: UK.

Van Loo, P., 2001b. Modulation of aggression in male mice: influence of cage cleaning regime and scent marks. In: Male Management: Coping with Aggression Problems in Male Laboratory Mice. Surrey: UK.

Van Loo, P., 2001c. Influence of cage enrichment on aggression behavior and physiological parameters in male mice. In: Male Management: Coping with Aggression Problems in Male Laboratory Mice. Surrey: UK.

Weed, J.L., Raber, J.M., 2005. Balancing animal research with the animal well-being: establishment of goals and harmonization. Inst. Lab. Anim. Res. J. 46 (2), 118—128.

Wolfle, T.L., 2005. Introduction: environmental enrichment. Inst. Lab. Anim. Res. J. 46 (2), 79—82.

Wurbel, H., Chapman, R., Rutland, C., 1998. Effect of feed and environmental enrichment on development of stereotypic wire-gnawing in laboratory mice. Appl. Anim. Behav. Sci. 60, 69—81.

Model Selection

6

D.L. Hickman[1], S. Putta[2], N.A. Johnston[1], K.D. Prongay[3]

Indiana University, Indianapolis, IN, United States[1]; University of Southern California, Los Angeles, CA, United States[2]; Oregon National Primate Research Center, Beaverton, OR, United States[3]

CHAPTER OUTLINE

6.1 SELECTION CRITERIA

When designing a research plan for the exploration of a scientific hypothesis, the scientist carefully evaluates all of the available models to ensure that they are selecting the most appropriate model. Although this text is focused on the use of animals in research, it is critical to recognize that animal models are only one category of model that is available to analyze biological processes. For example, in vitro models such as computer models and cell cultures are invaluable tools for the reduction of the overall number of animals that are used by providing the ability to determine potential toxicity before moving to live animals. The results of clinical observations and studies in humans can also provide scientists with useful information in their scientific evaluations. In general, scientists rely on in vitro tools in the early stages of their experimental evaluations and utilize in vivo animal models when they need to mimic the complicated interactions between cells, tissues, and organs that occur in all animals, including humans. This allows the scientist to better understand these interactions before working in the in vivo human model.

The selection of an appropriate animal model relies on a variety of criteria. At the most simple, animal models are used because they do something similar to humans, such as develop cancer, cardiovascular disease, or diabetes. Scientists then look for ways to prevent or treat these conditions that could be able to be applied to the equivalent human condition. Alternatively, they are selected because they do something that humans cannot do, for example, protect bone loss during hibernation. In this case, the scientist is hoping to identify the mechanism that creates that condition or ability to determine if it is something that can be applied to the human, such as using endocrine interventions to prevent osteoporosis in humans. Other advantages of animal models include the ability to control their environment, including diet, to minimize potential variables that may adversely affect the research and that they frequently have shorter life spans, allowing scientists to evaluate changes over the entire life of the animal. For example, in a mouse, a lifetime study is approximately 1.5−2 years, whereas for a human, the equivalent study would take over 70 years.

When selecting models that utilize animals, the ethical principles of the 3 R's (reduction, refinement, and replacement) (Russell and Burch, 1959) are applied, as mandated by federal regulations. The use of in vitro models embodies the ethical principle of replacement through the replacement of animals with nonanimal alternatives. When using animals, careful attention to the statistical design and experimental model can help ensure that the appropriate number of animals is used to obtain statistically relevant results without waste of animals, embodying the principle of reduction. Careful attention is also paid to the refinement of the actual techniques that are being utilized to ensure that they are as humane as possible. Implementation of the 3 R's into a study design is overseen by a variety of regulatory authorities worldwide. In the United States, specific regulations designate this oversight to local Institutional Animal Care and Use Committees, as described in Chapter 4.

6.2 TYPES OF MODELS

There is significant variety in the types of animal models that are available. The most desirable models are those that spontaneously occur in the animal, as these have the potential to provide the most information regarding how the condition develops naturally. However, there are not always spontaneous analogs in the animal kingdom, so scientists utilize a variety of other techniques to create their animal models. Models involving mice frequently are manipulated in ways to induce genetic changes—either through the induction of random mutations or the intentional manipulation of the mouse genome. Animals can also be exposed to various chemicals or surgical manipulation to induce a clinical condition. Behavioral evaluations are frequently utilized to evaluate the mental functioning of animal models of all species. All of these models will be discussed in further detail in this chapter.

6.3 SPONTANEOUS MUTATION MODELS

Naturally occurring or spontaneous diseases of animals allow us to gain significant fundamental knowledge of human disease and translational medicine. Spontaneous mutations are genetic events that happen randomly by chance. These mutations occur commonly in mice in many research colonies and are discovered as unexpected phenotypes that are abnormal. This observational approach is useful for spontaneous mutations that affect visible phenotypes such as coat color, size, morphology, and behavior. The discovery of these spontaneous mutations by vigilant animal care technicians and scientists who can rationally explore these chance observations can lead to novel mouse models for research. Despite the many advances in molecular biology and technologies that have enabled creation of mouse models using site-targeted mutagenesis, spontaneous mutations hold value and continue to provide valuable models to understand mammalian biology (Davisson, 2005).

Spontaneous mutations identify the novel genes and provide potential models for human inherited diseases for which the mutated gene has not yet been identified. Additionally, the genetic defects such as single base pair changes, deletions, and small insertions identified in most spontaneous mutations of humans and other mammals are confined to a single gene (Grompe et al., 1989). Also, some spontaneous mutations produce a phenotype that more or less closely resembles the human disorder than models generated via targeted mutations of the same gene. An example of this scenario is a spontaneous mutation of the homeobox D13 (*Hoxd13*) gene that closely resembled the human disorder of synpolydactyly (Dolle et al., 1993; Bruneau et al., 2001). The key steps required in the development of inherited spontaneous mutation model systems are discovery of these abnormal phenotypes and systematic analysis and capabilities to follow up these observations.

Spontaneous mutations occur during the cycle of DNA replication or repairing of damaged DNA. The average mutation rates are approximately 10^{-5} to 10^{-7} events per gene per generation (Schlager and Dickie, 1971; Melvold, 1975; Melvold et al., 1997). Thus the probability of identifying such events increases with increase in size of the breeding program; as well, the probability of finding specific recessive mutations is also greater with increased inbreeding. The spontaneous mutation rates also vary from one gene to another because the type of mutant genes found in this manner is biased by their phenotypic visibility and viability of the mouse with the mutation

(Davisson, 2005). Mutations with effects that are not obviously visible can be discovered by utilizing large scale screening techniques. To increase the rate of mutations so that screening can be efficient and cost-effective, researchers induce random mutations using various techniques.

6.4 RANDOM MUTATION MODELS

Mutations in mice can be generated by a plethora of techniques including those such as ionizing radiation or chemical agents. The induction of mutagenesis is a powerful tool to elucidate function of gene at single loci and discovering genetic contributors to complex disease phenotypes. The mutant mice being created from genome-wide mutagenesis projects worldwide are an invaluable resource for biomedical research community to study the molecular and physiological consequences of mutations at a whole organism level (Noveroske et al., 2000).

For radiation-induced deletions, males are exposed to ionizing radiation for various periods of time. They are next set up for breeding with nontreated females. The offspring, which will have inherited the mutations from their parents, are screened for phenotypes of interest.

The most commonly used chemical mutagen is N-ethyl-N-nitrosourea (chemical formula $C_3H_7N_3O_2$), also commonly known as ENU. This compound is a highly potent chemical mutagen that produces heritable mutations in spermatogonia, resulting in a range of mutant effects from complete or partial loss of function to exaggerated function, and discovers gene functions in an unbiased manner (Justice et al., 1999). After treatment of male mice with ENU, the mice are mated in genetic screens specifically designed to uncover mutations of interest.

6.5 GENETICALLY MANIPULATED MODELS

The development of genetic engineering technology combined with the availability of the complete sequence of the mouse genome has provided researchers with the tools to insert, delete, or alter virtually any desired gene. A number of different sophisticated gene-targeting strategies have been used to search for novel genes regulating disease phenotype and manipulate the mouse genome (Bradley and Liu, 1996). Broadly, these techniques can be organized into two categories: (1) transgenic mice produced using technique of introducing a functional exogenous gene randomly into the mouse genome (Hardouin and Nagy, 2000) and (2) mutant mice where a targeted endogenous gene is altered (for example, **knockout (KO) mice**, where gene expression is inhibited, and **knock-in mice**, where the expression of a typically silent gene is triggered).

Chimeric mouse generation is one technique utilized to produce genetically modified mice. In one technique, embryonic stem cells from one strain of mouse are aggregated with eight cell stage embryos. The resulting offspring are chimeras that can be screened for phenotypes of interest and cultivated using homologous recombination. These mice are not true transgenic animals, though, as the embryonic stem cells are donated from other mice. More recently, retrovirus-mediated gene transfer has been utilized as the vector to transfer genetic material into the developing cells. In both cases, the resulting chimera is selectively bred for the expression of the desired gene for at least 20 generations until homozygous offspring are born.

In DNA microinjection, the gene to be introduced is injected into the pronucleus of the recipient cell which is then cultured in vitro until it achieves the blastocyst stage. At that time, it is transferred to a recipient female mouse that carries the modified embryo to term. Offspring are screened for the gene of interest and selectively bred for the expression of the desired gene for at least 20 generations until homozygous offspring are born.

The use of genetically engineered mutant strains such as mutant phenotype may necessitate special husbandry. With transgenic mice, multiple copies of the transgene frequently insert at multiple sites, which might knock out endogenous genes creating a cryptic KO mouse. Also, expression of transgenes may be lost through breeding; and overexpression may have unintended consequences including cell lethality (Leiter et al., 2007). When developing and expanding a line of mutant mice, several issues must be clarified regardless of the origin of a random mutation such as heritability and mode of inheritance of the phenotype, genetic background of the new mutant strain as well as breeding scheme to expand and maintain mutant line to minimize any genetic drift (Hardouin and Nagy, 2000).

6.6 CHEMICALLY INDUCED MODELS

Chemically induced animal models of human disease are widely used by biomedical researchers (examples include Blandini and Armentero, 2012; King, 2012; Gao and Zheng, 2014). This type of model may be categorized into two broad groups, which we will briefly discuss in this section.

In the first category, animals are induced to develop a pathology or disease to study that pathology or disease. The model development is artificial, but the end result is a usable model of the human disease condition. One example of this is the study of diabetes mellitus through the administration of streptozotocin (STZ), a compound that causes specific necrosis of the pancreatic beta cells. Although the rat is most commonly used as a model of diabetes created with STZ, other species, such as pigs, nonhuman primates, and chickens have also been used. The disease process in humans is multifactorial, caused by insulin deficiency or insulin resistance, and progresses with an unpredictable onset. By contrast, protocols for inducing diabetes with STZ involve as few as one dose of the compound. Human diabetes patients rarely suffer beta-cell destruction in this way, but the end result in animals after STZ administration mimics the human condition well enough to provide meaningful data in the field of diabetes research. New variations of this model have been developed, such as adding a high fat diet to the STZ to more closely mimic human diabetes (Skovsø, 2014). Similarly, chemically induced osteoarthritis (OA) is used widely to study the disease condition and possible therapies (Suokas et al., 2014). Intraarticular injection of monoiodoacetate (MIA) is the common chemically induced version of this model (Suokas et al., 2014). The MIA is injected into the joint, where it targets the articular chondrocytes, and causes eventual cell death (Ferreira-Gomes et al., 2012). This progressive loss of chondrocytes, destruction of the joint, and resulting pain mimic the natural progression of OA in humans (Guzman et al., 2003). The human pathogenesis of OA has not been definitively defined, but several risk factors have been identified, including aging, obesity, and genetic predisposition (Xia et al., 2014). The animal model is also used to study pain in general and possible therapies for pain, since the resulting damaged joint results in significant pain with movement.

In the second broad category, generally encompassing toxicology, the animal model is created in a way to mimic human exposure to a chemical and the resulting disease progression. As it is not ethical to conduct preliminary toxicology testing on humans, in vivo models must be used, and data extrapolated to human physiology. Using an animal model is one of the best ways to see the effects of a chemical in the complex and interconnected biology of a living being. Toxicology studies work under the assumption that the animal selected for the study will react in the same way as a human would. As different species have different reactions to compounds, most safety testing is performed on several species in the effort to capture as much information about possible toxic effects as possible. For approval by the US Food and Drug Administration (FDA), the typical compound is tested in both a rodent (usually rat) and a nonrodent. Some of the limitations with animal models can be overcome by using transgenic and/or humanized animals. These animals may be better able to mimic a human response to the test compounds (Stanley, 2014) than "regular" animals, as the genetically modified humanized animals have been created to better mimic the human immune system response or physiology (Boverhof et al., 2011; Cohen, 2014). Animal models are used to assess both acute and chronic toxic effects of the test chemicals. In chronic studies, animals are assessed for neurologic, reproductive, and carcinogenic changes for weeks to months after administration of the chemical. Acute responses to a chemical are assessed between hours and days after administration. More information on the study of toxicology can be found in other references (Le Floc'h et al., 2012; Gad, 2014).

6.7 PHYSICALLY INDUCED MODELS

Trauma is the leading cause of death for Americans between 1 and 44 years old, and it is the third most common cause of death overall. Each year, traumatic injury accounts for 41 million emergency department visits and 2 million hospital admissions. The economic impact is greater than $406 billion per year (Institute, 2015).

War, automobile injuries, burns, falls, and other accidents are devastating for the individual recipients of trauma and those who care for them (Humphreys et al., 2013; Rowe et al., 2013). Medical research in the field of traumatic injury treatment has been responsible for improving skills, treatments, and equipment used in treating humans. Physical trauma is often modeled with animals, as no other system could mimic the multiple systems/whole body nature of traumatic injury. As trauma is rarely limited to one organ or pathogen, the use of animal models to mimic the complex nature of the problem is essential for conducting research on treatments and therapies. The small and large animal model development has been identified as critical to advancing the field of trauma care (Valparaiso et al., 2015). This section will provide a few examples of commonly used animal models of traumatic injury.

Humane considerations to the animals used as models in these studies are given the utmost importance. Animal models are created to achieve a balance between mimicking the pathology and minimizing the animal suffering. This goal can best be accomplished with precise methods that standardize the trauma, localized and limited to specific areas on the body that correspond to the injury to be studied, and the use of agents, both pharmaceutical and physical, to support the animal in this process. The use of animal models has been essential for the discovery of advancements in this field of research. The use of these animal models must be justified scientifically and ethically given the degree of potential pain and distress. Supportive care to the animals helps meet this balance.

Traumatic brain injury (TBI) is a major cause of long-term disability worldwide (Humphreys et al., 2013; Malkesman et al., 2013; Angoa-Perez et al., 2014; Jean et al., 2014; Petraglia et al., 2014). TBI is one of the most common neurological diagnoses in the United States, with an estimated 1.5 million people sustain TBI on a yearly basis (Angoa-Perez et al., 2014). Both mice and rats are widely used as models as they are available in numerous genetically altered strains. Disadvantages to the rodent model include great physiologic and anatomic differences between the human and rodent brain, and the inability to model some of the cognitive function and emotional changes in humans in the long-term sequelae of TBI. Certainly, the physiological and anatomical changes in the CNS are able to be studied in animal models. Inflammatory changes, the immune system reaction, and other neuropathological changes can be measured and studied using an animal model system, which is not possible with any other method. Although the same cognitive assessments performed in humans are not possible, many of the established behavioral tests designed for rodents can be used after TBI to assess for depression-like symptoms, anxiety-like symptoms, and aggression (Malkesman et al., 2013), well-known long-term signs of TBI.

The cause, duration, and severity of TBI in humans vary considerably. TBI associated with war-related injuries may be as a result of a single, severe impact to the head. Mild but repeated TBI may be caused by sports injuries acquired in football or boxing (Malkesman et al., 2013). Animal models have been created to mimic these different degrees of TBI. Various protocols to create TBI are described in the literature, including direct impact on the skull by a set weight, lateral fluid percussion injury, and blast-induced injury generated by a pressurized gas (Petraglia et al., 2014). All studies must be performed in a precise and controlled method to generate the same degree and severity of injury in each animal in the study. All animals in a group should respond in the same way to the injury if valid results are to be generated in treatment or prevention experimental groups.

Traumatic injury due to burns is a significant human health problem. Burns account for an estimated 265,000 deaths worldwide. In the United States, burn injuries account for about 450,000 cases of medical treatment (American Burn Association, 2015). Animal models provide a method for reproducing the complex pathophysiology that occurs in burn wounds. Three species are commonly used for burn research: pig, mouse, and rat. The pig model is often considered the most relevant to the human condition. Pig skin is more similar in anatomy and physiology than rodents (Branski et al., 2008; Abdullahi et al., 2014; Sheu et al., 2014), rendering the research more clinically relevant. Burns can be created on the skin using a heated metal rod applied to the skin, or exposing an area of skin to hot water, simulating a scald (Abdullahi et al., 2014; Sheu et al., 2014). The severity of the burns will depend on the temperature of the rod or water and the duration of exposure. Burn wounds have been categorized into first, second, or third degree based on the depth and extent of the damage. Burn creation protocols using animal models have been standardized for time and temperature to accurately mimic the severity of burn to be studied (Branski et al., 2008; Venter et al., 2014). As with all animal studies that might cause significant pain or distress, pain-relieving drugs should be administered whenever compatible with the scientific objectives of the study.

Noise-induced hearing loss (acoustic trauma) is the most widespread disability in modern society (Pfannenstiel, 2014; Stucken and Hong, 2014). The source of damaging noise may be through occupational settings (e.g., factory), military activities (e.g., explosions, firearms), or recreational activities (e.g., firearms, snowmobiles). Animals have been used to mimic the damage to the auditory system to better study the pathology of noise-induced hearing loss and work on preventative and therapeutic options. Common animal models include mice, rats, and chinchillas. Rat models are widely used, as rats can be trained to demonstrate a behavior in response to sound exposure (Turner et al., 2005; Hayes et al., 2014; von der Behrens, 2014). The hearing abilities of these trained animals can be assessed without anesthesia or invasive procedures. Chinchillas are a common animal used in hearing-related studies, as the hearing range of these animals mimics the human range more closely than other animals (Heffner and Heffner, 1991, 2007). In addition, the auditory structures are large and easily accessible, facilitating any procedures or manipulation. In addition to the traditional animal models, nonvertebrate animals have been recently proposed as models to study noise-induced hearing loss. Sea anemones and fruit flies are successful models for studying certain aspects of acoustic trauma (Christie and Eberl, 2014). Mechanosensory structures of sea anemones and fruit flies are similar in function to human auditory hair cells. The fundamental similarities make these animals useful models to supplement the more traditional methods of studying hearing loss.

Interestingly, animals can act as models of traumatic disease outside of the laboratory setting. In veterinary clinical practice, trauma to animals is a common medical emergency, accounting for 11—13% of all animals examined in urban veterinary teaching hospitals (Streeter et al., 2009; Hall et al., 2014). Dogs sustain severe traumatic injuries from cars or other vehicles. These injuries are similar to human trauma patients, lending these cases to serve as clinical models for human trauma (Hall et al., 2014). This type of model is much different from the homogenous population of rodents receiving precise, controlled identical procedures to create the trauma. However, the clinical canine model can be seen as a useful tool for translational research for interventional trails of human therapies prior to human clinical trials. Taken together, both models have the ability to advance knowledge on more effective prevention and treatment of traumatic injuries.

6.8 SURGICAL MODELS

Surgical models are used to approximate a variety of clinical conditions. There are generally three categories of surgical models. The first induces a pathologic change, but the development of the pathologic change may not mirror that seen in the human pathology. These models are generally of interest to scientists studying disease treatment. A second category of surgical models involves the development of novel surgical techniques that can be used in humans. A third category involves the use of animal models as an adjunct training model for human and animal surgeons. In this section, we will briefly review some of the more common surgical models.

6.8.1 EXAMPLES OF CARDIOVASCULAR AND THORACIC SURGICAL TECHNIQUES

Studies requiring multiple blood samples or prolonged sampling or infusion typically utilize surgically placed vascular catheters. Venous catheters are used to collect venous blood samples and to infuse drugs and test substances. Arterial catheters may be used for blood gas sampling or direct measurement of blood pressure. The vessel and catheter designs vary by species and experimental need. Following catheter placement, the catheter is tunneled through subcutaneous tissue to a convenient collection site such as the dorsal cervical region in small animals or the proximal flank or dorsum in sheep, dogs, and other large species (Wingfield et al., 1974).

Cardiac catheterization is used to measure coronary arterial blood flow and pressure in the cardiac chambers. Various dyes can be infused through the catheter for coronary angiography or left ventricular angiography. Catheters are inserted into either the femoral artery and vein, or brachial artery and vein, and advanced into the coronary artery or left ventricle. In small animals, the cardiac catheter is placed in the carotid artery and used to measure systolic and diastolic blood pressure (Cristina Lorenzo et al., 2009). When advanced into the left ventricular chamber, the catheter may also be used to evaluate left ventricular function in systole and diastole.

Other common thoracic surgical techniques include coronary artery bypass (Shofti et al., 2004), cardiopulmonary bypass (DiVincenti et al., 2014), and pulmonary autographs (Nappi et al., 2014). A number of species, including mice, rats, dogs, and sheep are used for cardiac transplant (Cooper, 2012a). Pig models of experimental cardiac repair techniques for valve replacement include a biodegradable ring for mitral and tricuspid annuloplasty (Kalangos et al., 2006; Myers and Kalangos, 2013). Tissue-engineered vessels are implanted into small animals to evaluate patency and evaluate "mechanistic hypothesis" (Swartz and Andreadis, 2013). Large animal models, particularly sheep and dogs, are used to assess long-term patency, remodeling following graft placement, and to evaluate function over time. Tracheostomies are most frequently used therapeutically, following mechanical obstruction of the trachea with mucus, but may also be used experimentally, for administering gas or volatile substances (Alfredo and Laura, 2009).

6.8.2 EXAMPLES OF GASTROINTESTINAL SURGICAL TECHNIQUES

Surgical models of acute periodontal defects are generated by surgical removal of the bone, cementum, and periodontal ligament. Chronic models are created by placing a foreign body, such as orthodontic elastic or sutures around a tooth (Struillou et al., 2010). While mice, rats, ferrets, mink, and other species are used in periodontal research, the dentition and mastication of many small animals differ significantly from humans. Nonhuman primates and dogs are the most commonly used species for these studies (d'Apuzzo et al., 2013).

When studying the function of the gastrointestinal tract, there are multiple surgical approaches that can be utilized. Cannulas for sample collection or infusing can be placed in the stomach and small intestine (Hill et al., 1996; Alfredo, 2009). Evaluation of the physiology and pathology of specific intestinal segments is achieved through a variety of techniques such as **intestinal loop isolation** (Anderson et al., 1962). Models of polymicrobial experimental sepsis are generated using the **cecal ligation and puncture** technique. Here, the cecum is perforated, releasing material from the digestive tract into the peritoneal cavity (Toscano et al., 2011).

Samples of hepatic tissue may be collected percutaneously with a biopsy needle (Voss, 1970; Varagona et al., 1991). If larger samples are required, a cranial midline laparotomy or laparoscopic biopsy is performed (Rawlings et al., 2000). **Liver transplantation** in the mouse, rat, and pig has been used to evaluate surgical techniques, transplant immunobiology, and to identify postsurgical transplant issues (Tanaka et al., 1993; Chen et al., 2013; Nagai et al., 2013). Bile may be collected directly by cannulation of the gall bladder, and the common bile duct may be cannulated in larger species (Boegli and Hall, 1969).

6.8.3 EXAMPLES OF URINARY SYSTEM TECHNIQUES

Chronic renal failure is generated using the *remnant kidney model*. In this model, one kidney is removed and two-thirds of the blood flow to the remaining kidney is obstructed. The model is used to evaluate the pathophysiology of chronic kidney disease in dogs and rats (Brown, 2013). Models of acute renal failure are generated by **renal clamping** (Golriz et al., 2012). Renal ischemia is produced by placing a clamp on the renal artery. Removal of the clamp produces reperfusion injury (Wei and Dong, 2012). The mouse model of kidney transplantation was first described in 1973. Initially, the model was used to assess survival posttransplant. More recent studies have grafted a kidney into a recipient mouse. The model has been used to assess acute rejection, cellular and humoral rejection mechanisms, and treatment modalities. Pigs are a common large animal (Golriz et al., 2012).

6.8.4 EXAMPLES OF ORTHOPEDIC TECHNIQUES

Surgical induction of osteoarthritis is achieved by altering the exerted strain on the joint or altered load bearing with the resulting instability causing osteoarthritis (Lampropoulou-Adamidou et al., 2014). Mice, rats, and large animals have also been used to evaluate the design and surgical placement of orthopedic implants. With the increasing number of joint replacement surgeries, these models have been used to assess bacterial infection and to investigate both novel antimicrobial therapies and novel application techniques (Stavrakis et al., 2013), as well as evaluation of long-term implant failure (Cordova et al., 2014).

Spinal cord regeneration studies utilize direct surgical transection or partial transection of the spinal cord (Cheriyan et al., 2014). Models of scoliosis can be surgically induced in both quadrupedal and bipedal types. Rodent models may be better suited for evaluating corrective interventions than evaluating the causes of scoliosis in humans (Roth et al., 2013), while larger animals have been studied using a posterior spinal tether or compression-based fusionless scoliosis correction device (Bobyn et al., 2014).

6.8.5 EXAMPLES OF REPRODUCTIVE SYSTEM TECHNIQUES

Ablation of the ovaries is used to generate ovarian hormone deficiency in a wide variety of disciplines (Wronski et al., 1985). In the rat model of menopause, ovary ablation results in a hormone imbalance that causes an increase in bone resorption relative to formation, resulting in osteopenia.

Pigs, sheep, and nonhuman primates are the most common models used for intrauterine fetal surgery. Much early work to develop techniques for repair of cardiac abnormalities in the developing fetus was done in sheep (Hoffman et al., 1996). Baboon and rhesus macaque models of preterm birth require instrumentation of the preterm fetus and caesarian section delivery (Grigsby et al., 2012).

6.8.6 EXAMPLES OF ENDOCRINE SYSTEM TECHNIQUES

Surgical induction of diabetes is achieved by **pancreatectomy**. Removal of 95% of the pancreas induces diabetes in 3 months in rat models, and a similar model is available in dogs and swine (Sarr, 1988). **In situ perfusion of the pancreas** is used to evaluate pancreatic hormones such as insulin. A perfusion catheter is placed in the celiac artery and samples are collected through a second catheter at the portal vein. Thyroidectomy, thyroparathyroidectomy, and parathyroidectomy are used to induce hypoparathyroidism, to evaluate renal management of phosphate, to modify bone remodeling, and to study the effect of parathyroid hormone with an intact thyroid.

6.8.7 EXAMPLES OF DERMATOLOGIC TECHNIQUES

A variety of infusion and sampling ports can be surgically implanted in the subcutaneous tissue. The devices may connect to structures such as arteries or veins, or be connected directly to a reservoir. Access ports should be of an appropriate size for the species and sample, and have been successfully used in a variety of species. Infusion of drugs or test vehicles may be achieved by placement of an osmotic mini-pump (Alzet osmotic pump, ALZA Corporation, Palo Alto, California). These devices provide a specific fluid volume over a defined time period.

6.8.8 EXAMPLES OF NEUROLOGICAL TECHNIQUES

A stereotaxic device uses a set of three coordinates that, when the head is in a fixed position, allows for the precise location of brain sections. Stereotactic surgery may be used to implant substances such as drugs or hormones into the brain. Electrodes, for monitoring response or providing controlled stimulus, may also be implanted. When such implants are used for a period of time, a head plate may be attached to the skull to protect the implant (Gardiner and Toth, 1999).

Surgically induced stroke models are grouped into three broad methods, namely, *embolism*, *middle cerebral artery occlusion*, and *perforating artery occlusion*. Mice and rats are most commonly used (Casals et al., 2011).

6.9 BEHAVIORAL MODELS

Animal behavior models are used to evaluate neuropsychiatric disorders such as addiction, behavior disorders such as depression, the effects of infection on brain function and development, the impact of pharmacologic and toxic agents, and to study age-related cognitive loss. The validity of these models is classically assessed using criteria proposed in 1969 by McKinney and Bunney (McKinney and Bunney, 1969). They suggested that animal behavioral models should resemble the etiology, biochemistry, symptoms, and treatment of the neurobehavioral disease being studied. Additionally, the models should be reproducible, reliable, predictive, generalizable, and relevant (McKinney and Bunney, 1969). To achieve this, behavioral models are initially developed in one species, typically a rat or mouse. Variables such as housing, diet, gender, age, and the type of test administered are limited. Only after confirming replication and reproducibility, the model is expanded to evaluate the effect of different housing systems, different sexes and ages, and using multiple tests to objectively assess the same trait (van der Staay et al., 2009).

Behavioral models can be categorized into two broad classes: naturally occurring disease/defect models and experimentally induced models. Each is used to evaluate how neuropathways control behavior and to assess the impact on these pathways of mechanical or pharmacologic disruption or administration. The information from these studies provides insights into normal and abnormal behavior physiology and pathology, and to develop targets, pathways, and mechanism of drug action and behavioral modification (van der Staay et al., 2009).

Normal models are used to develop a comprehensive understanding of normal behavior and neurobiology. An accurate assessment of normal behavior is essential for recognition and evaluation of abnormal behavior. Classically, these assessments may be divided into behavioral tests and tests of learning and memory. Behavioral tests evaluate an animal's response to a situation. The open field test, the elevated plus maze test (Fig. 6.1), the acoustic startle test, and the intruder test are common examples (Van Meer and Raber, 2005). Tests of learning and memory rely on an animal's ability to remember an object or pattern and include object recognition, object

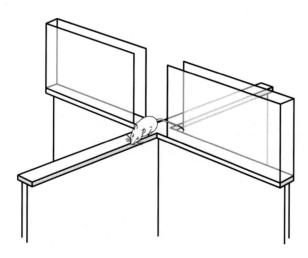

FIGURE 6.1

Schematic of elevated plus maze. The amount of time the rodent is willing to spend in the open arms is reflective of how willing the rodent is to explore its surroundings, providing a reflection of the amount of anxiety being expressed by the animal (Mulder and Pritchett, 2004).

permanence, the water maze (Fig. 6.2), and variety of pattern memorization tests (Kalin and Shelton, 1989). The results from normal animals are compared to those with spontaneous or induced neural deficits to better understand neurologic functioning.

6.9.1 NATURALLY OCCURRING DISEASE MODELS

Naturally occurring disease models are used to study the effects of age on cognition and naturally occurring infectious disease; and they are used in selective breeding for malformations and congenital disorders or in breeding to enhance behavioral extremes (van der Staay et al., 2009). Models of age-related changes exist in a number of species, but the rat, mouse, and nonhuman primate are most commonly used. Cognitive decline of rhesus macaques is similar to that observed in humans. Histologic changes have been tied to this cognitive decline (Hof et al., 2000; Peters et al., 2008).

Naturally occurring infectious disease models include a number of viral, bacterial, and fungal diseases that infect the CNS. Generally, these models are used to

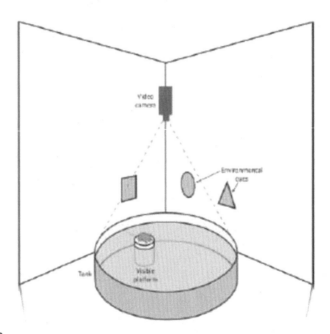

FIGURE 6.2

Schematic of Morris water maze. The small rodent is placed in the tank at the insertion point. Visual landmarks are provided to allow the small rodent to navigate to the platform that is located just below the surface of the water. By performing repeated trials, this test is a measurement of learning and memory (Mulder and Pritchett, 2003).

evaluate the clinical course of a disease and to identify and monitor the neurobehavioral changes caused by neurodegeneration, dystrophic mineralization, and similar inflammation-related pathologies. Full-term fetuses from rhesus macaques infected with cytomegalovirus provide a model of mental retardation and sensorineural hearing loss (London et al., 1986; Tarantal et al., 1998).

Malformations and congenital disorders may result from spontaneous mutations and offer unique models. The mouse model of cerebellar dysfunction and the primate model of neurocutaneous melanosis are examples (Lalonde and Strazielle, 2007; Chen et al., 2009).

6.9.2 EXPERIMENTALLY INDUCED MODELS

Experimentally induced deficits comprise a large variety of models and species. These include environmentally induced models to evaluate the effect of acute and chronic pain, stress, or sleep deprivation; dietary induced models such as tryptophan depletion, hypoxia, and anorexia; drug-induced behavioral models; stroke models; and models of infection and inflammation. Three of the most common model types are addiction, depression, and teratogen exposure, and these are discussed later.

Some animal models mimic various aspects of drug and alcohol addiction. Some models are used to evaluate the neurobiological basis of addiction. Others are designed to assess what role behavioral reinforcement plays in the initiation, maintenance, and reinstatement of drug and alcohol addiction. Broadly, they may be used to assess aspects of addiction behavior as divided into four categories. **Positive reinforcement effects models** evaluate the role of positive reinforcement in the development of addiction. These models use either intravenous or oral drug administration and may include an operant conditioning component to assess the level of motivation and discrimination components to assess preference (Shippenberg and Koob, 2002). **Drug withdraw models** are used to assess both the markers of withdrawal and evaluate motivation (Shippenberg and Koob, 2002). These models have demonstrated, for example, that rats allowed access to a 10% alcohol solution during withdrawal will preferentially lever press, maintaining a steady blood alcohol level (Roberts et al., 1996). **Escalation of intake models** are an operant conditioning model of excessive drug intake and may be used to model uncontrollable drug-taking and drug-seeking behaviors (Edwards and Koob, 2013). **Relapse models** are used to model contextual and stress stimuli that reinforce relapse (Weiss, 2010).

Toxicity and teratogen models evaluate the effect of a drug or teratogen on fetal development. Dams may be exposed to the substance throughout pregnancy or during a particular phase of fetal development (Lucey and Behrman, 1963). Offspring are evaluated for cognitive development and behavior (Kietzman et al., 2014).

6.10 PRODUCTION MODELS

The concept of using animals for the production of some usable commodity is very familiar in the agriculture sector. Animals have been used for meat, egg, milk, and fiber production for thousands of years. In the area of biomedical science, animals are also used to produce a variety of substances for research or medical use. Animals are capable of producing complex, biologically active proteins easily and efficiently (Wang et al., 2013). The use of animals in the production of monoclonal and polyclonal antibodies is essential (Council, 1999). Antibody use in biomedical science and research is essential to many types of research, diagnostic, and therapeutic procedures. Advancing technology in the production of transgenic animals has also made it possible for such animals to produce recombinant proteins in milk. These foreign proteins in the milk can be extracted and purified for many purposes, such as pharmaceuticals, medical therapies, or even biomaterials. When considering animals as production models, one must consider the use of transgenic large animals for the production of replacement organs for xenotransplantation, commonly defined as living cells, tissues, or organs of animal origin transplanted into a recipient of another species, such as a human. Given the limited supply of human organs for donation and the high demand for such transplants, xenogeneic materials have the potential to meet this need to improve human health.

The ability of an intact systemic animal immune system to produce antibodies is capitalized by scientists to produce specific antibodies to selected antigens in diagnostic and research applications. The use and application of antibodies is essential for work in experimental biology, medicine, biomedical research, diagnostic testing, and therapy (Leenaars and Hendriksen, 2005). The use of animals in the production of antibodies is essential. More detailed information on antibody production methods can be found in other references (Hanly et al., 1995; Council, 1999). The method of producing antibodies involves preparation of the specific antigen and injecting this antigen into the appropriate species of animal to generate an immune response. The antibodies are produced by B-lymphocytes and circulate throughout the blood and lymph. To produce monoclonal antibodies, the immunized animal is euthanized and the spleen removed. A single clone of the correct B-lymphocyte is fused with an immortal myeloma cell, resulting in a hybridoma cell that can produce a virtually unlimited supply of monoclonal antibodies. In this production strategy, mice are typically used as the initial immunized animal since the myeloma cells available for fusion with the B-lymphocyte are murine origin. The propagation of the hybridized cells can occur either in vivo or in vitro. In vivo, the hybridoma cells are injected into the intraperitoneal cavity of a mouse and the resulting ascites fluid contains the desired antibody. Because this technique raises concerns about the humane use of animals, as the development of ascites may result in pain and distress, alternative in vitro methods and more humane protocols are often considered (Leenaars and Hendriksen, 2005).

The generation of polyclonal antibodies can be performed in a variety of animal species. The selection of the animal species is determined by the needs of the antibody production: the amount needed, the ease of obtaining blood samples, the relationship between the antigen and the animal species, and the intended use of the polyclonal antibody. The greater the phylogenetic distance between the source of the antigen and the species of the animal to be immunized, the more robust the immune response will be, as the difference in the amino acid sequences between the proteins will be greater. The antibodies are produced by B-lymphocytes and circulate in the blood and lymph. To collect the antibodies, blood is withdrawn and purification methods extract the desired antibodies. The selection of which species of animal for this production is critical to the success for the intended use and quantity of the antibody. The animal must be large enough for easy and plentiful blood collection, and it must be able to mount a robust immune response to the antigen. Many species of animals have been used, including the mouse, hamster, rat, guinea pig, rabbit, goat, sheep, horse, and chicken (Hanly et al., 1995).

In addition to antibody production, animals can produce other substances required for biomedical research, therapies, or scientific applications. Large animals such as sheep, goats, and cattle are routinely used for milk production for human consumption. The mammary gland is an ideal bioreactor, and the expression of foreign proteins in the milk is relatively easy to accomplish (Wang et al., 2013). Transgenic goats, pigs, and rabbits have been created to produce milk containing proteins specifically altered for therapeutic treatment or generation of novel biomaterials (Williams, 2003; Tokareva et al., 2013). For example, transgenic goats have been created to express human immunocompounds in milk to better supplement the diet of children suffering from diarrhea and malnourishment (Bertolini et al., 2014). The advantages to this technique are numerous. The collection of milk from animals is easily performed. The foreign proteins produced in the mammary gland can be produced at rates of several grams per liter of milk, making this an efficient production system (Wang et al., 2013). Examples of proteins produced in milk for human use include human serum albumin, human clotting factor IX, and antithrombin III.

Currently in the United States, many people are on waiting lists for organ transplantations as a surgical solution to organ failure. However, there is only one registered organ donor per every ten patients waiting for transplantation. Even with improvements in medical care and surgical techniques for people who need organ transplants, the limited supply of organs is the limiting factor (Zeyland et al., 2015). One alternative to human organ donation is to use organs from animals, also known as xenotransplantation. This concept is not new (Cooper, 2012a), as physicians have tried to transplant tissues and organs from animals for over 300 years. Currently in common use, xenotransplantation of certain tissues, for example, porcine heart valves (Manji et al., 2012) and porcine corneas (Kim et al., 2013, 2014; Cohen et al., 2014; Lee et al., 2014), is successful. Solid organ transplantation is not currently possible, but advancements in the field of immunology are closing the gap and bringing the hope that transgenic animals will provide a source of needed organs. Current research has focused attention on the creation of transgenic pigs to supply solid organs for human transplantation. Hurdles to using pigs include the human recipient immune response to the foreign tissue and subsequent rejection. Pigs created for the purpose of organ donation would need to be multitransgenic to be compatible with the human immune system (Zeyland et al., 2015). The list of currently available transgenic strains of pig available for xenotransplantation research is lengthy, and many researchers believe that the barriers to mass produced organs for transplantation will be overcome (Cooper, 2012a,b).

REFERENCES

Abdullahi, A., Amini-Nik, S., Jeschke, M.G., 2014. Animal models in burn research. Cell. Mol. Life Sci. 71, 3241–3255.

Alfredo, R., 2009. In situ isolation of the stomach. In: Experimental Surgical Models in the Laboratory Rat. CRC Press.

Alfredo, R., Laura, I.P., 2009. Tracheostomy. Experimental Surgical Models in the Laboratory Rat. CRC Press.

Anderson, R.E., Diffenbaugh, W.G., Schmidtke, W.H., 1962. Isolated circular intestinal loop. Loop with an access fistula for experimental studies of intestinal function. Arch. Surg. 84, 559–563.

Angoa-Perez, M., Kane, M.J., Briggs, D.I., Herrera-Mundo, N., Viano, D.C., Kuhn, D.M., 2014. Animal models of sports-related head injury: bridging the gap between pre-clinical research and clinical reality. J. Neurochem. 129, 916–931.

American Burn Association, 2015. Available: http://www.ameriburn.org/index.php.

Bertolini, L., Bertolini, M., Murray, J., Maga, E., 2014. Transgenic animal models for the production of human immunocompounds in milk to prevent diarrhea, malnourishment and child mortality: perspectives for the Brazilian Semi-Arid region. BMC Proc. 8, O30.

Blandini, F., Armentero, M.T., 2012. Animal models of Parkinson's disease. FEBS J. 279, 1156–1166.

Bobyn, J.D., Little, D.G., Gray, R., Schindeler, A., 2014. Animal models of scoliosis. J. Orthop. Res. 33 (4), 458–467.

Boegli, R., Hall, I., 1969. A surgical external biliary fistula for the total collection of bile from rabbits. Lab. Anim. Care 19, 657–658.

Boverhof, D.R., Chamberlain, M.P., Elcombe, C.R., Gonzalez, F.J., Heflich, R.H., Hernández, L.G., Jacobs, A.C., Jacobson-Kram, D., Luijten, M., Maggi, A., Manjanatha, M.G., Benthem, J.V., Gollapudi, B.B., 2011. Transgenic animal models in toxicology: historical perspectives and future outlook. Toxicol. Sci. 121, 207–233.

Bradley, A., Liu, P., 1996. Target practice in transgenics. Nat. Genet. 14, 121–123.

Branski, L.K., Mittermayr, R., Herndon, D.N., Norbury, W.B., Masters, O.E., Hofmann, M., Traber, D.L., Redl, H., Jeschke, M.G., 2008. A porcine model of full-thickness burn, excision and skin autografting. Burns 34, 1119–1127.

Brown, S.A., 2013. Renal pathophysiology: lessons learned from the canine remnant kidney model. J. Vet. Emerg. Crit. Care (San Antonio) 23, 115–121.

Bruneau, S., Johnson, K.R., Yamamoto, M., Kuroiwa, A., Duboule, D., 2001. The mouse Hoxd13(spdh) mutation, a polyalanine expansion similar to human type II synpolydactyly (SPD), disrupts the function but not the expression of other Hoxd genes. Dev. Biol. 237, 345–353.

Casals, J.B., Pieri, N.C., Feitosa, M.L., Ercolin, A.C., Roballo, K.C., Barreto, R.S., Bressan, F.F., Martins, D.S., Miglino, M.A., Ambrosio, C.E., 2011. The use of animal models for stroke research: a review. Comp. Med. 61, 305–313.

Chen, J., Gong, W., Ge, F., Huang, T., Wu, D., Liang, T., 2013. A review of various techniques of mouse liver transplantation. Transplant. Proc. 45, 2517–2521.

Chen, Y., Deng, W., Zhu, H., Li, J., Xu, Y., Dai, X., Jia, C., Kong, Q., Huang, L., Liu, Y., Ma, C., Xiao, C., Liu, Y., Li, Q., Bezard, E., Qin, C., 2009. The pathologic features of neurocutaneous melanosis in a cynomolgus macaque. Vet. Pathol. 46, 773–775.

Cheriyan, T., Ryan, D.J., Weinreb, J.H., Cheriyan, J., Paul, J.C., LaFage, V., Kirsch, T., Errico, T.J., 2014. Spinal cord injury models: a review. Spinal Cord 52, 588–595.

Christie, K.W., Eberl, D.F., 2014. Noise-induced hearing loss: new animal models. Curr. Opin. Otolaryngol. Head Neck Surg. 22, 374–383.

Cohen, D., Miyagawa, Y., Mehra, R., Lee, W., Isse, K., Long, C., Ayares, D.L., Cooper, D.K., Hara, H., 2014. Distribution of non-gal antigens in pig cornea: relevance to corneal xenotransplantation. Cornea 33, 390–397.

Cohen, J., 2014. Toxicology. 'Humanized' mouse detects deadly drug side effects. Science 344, 244–245.

Cooper, D.K., 2012a. A brief history of cross-species organ transplantation. Proc. (Bayl. Univ. Med. Cent.) 25, 49–57.

Cooper, D.K., 2012b. The value of xenotransplantation. Bull. Am. Coll. Surg. 97, 54.

Cordova, L.A., Stresing, V., Gobin, B., Rosset, P., Passuti, N., Gouin, F., Trichet, V., Layrolle, P., Heymann, D., 2014. Orthopaedic implant failure: aseptic implant loosening—the contribution and future challenges of mouse models in translational research. Clin. Sci. (Lond.) 127, 277–293.

Council, N.R., 1999. Monoclonal Antibody Production. The National Academies Press, Washington, DC.

Cristina Lorenzo, C.N., Manuel, R.G., Laura, K., Martin, D., 2009. Cardiac Catheterization. Experimental Surgical Models in the Laboratory Rat. CRC Press.

d'Apuzzo, F., Cappabianca, S., Ciavarella, D., Monsurro, A., Silvestrini-Biavati, A., Perillo, L., 2013. Biomarkers of periodontal tissue remodeling during orthodontic tooth movement in mice and men: overview and clinical relevance. ScientificWorldJournal 2013, 105873.

Davisson, M.T., 2005. Discovery genetics: serendipity in basic research. ILAR J. 46, 338–345.

DiVincenti Jr., L., Westcott, R., Lee, C., 2014. Sheep (*Ovis aries*) as a model for cardiovascular surgery and management before, during, and after cardiopulmonary bypass. J. Am. Assoc. Lab. Anim. Sci. 53, 439–448.

Dolle, P., Dierich, A., Lemeur, M., Schimmang, T., Schuhbaur, B., Chambon, P., Duboule, D., 1993. Disruption of the Hoxd-13 gene induces localized heterochrony leading to mice with neotenic limbs. Cell 75, 431–441.

Edwards, S., Koob, G.F., 2013. Escalation of drug self-administration as a hallmark of persistent addiction liability. Behav. Pharmacol. 24, 356–362.

Ferreira-Gomes, J., Adaes, S., Sousa, R.M., Mendonca, M., Castro-Lopes, J.M., 2012. Dose-dependent expression of neuronal injury markers during experimental osteoarthritis induced by monoiodoacetate in the rat. Mol. Pain 8, 50.

Gad, S.C., 2014. Animal Models in Toxicology, second ed. CRC Press.

Gao, F., Zheng, Z.M., 2014. Animal models of diabetic neuropathic pain. Exp. Clin. Endocrinol. Diabetes 122, 100–106.

Gardiner, T.W., Toth, L.A., 1999. Stereotactic surgery and long-term maintenance of cranial implants in research animals. Contemp. Top. Lab. Anim. Sci. 38, 56–63.

Golriz, M., Fonouni, H., Nickkholgh, A., Hafezi, M., Garoussi, C., Mehrabi, A., 2012. Pig kidney transplantation: an up-to-date guideline. Eur. Surg. Res. 49, 121–129.

Grigsby, P.L., Novy, M.J., Sadowsky, D.W., Morgan, T.K., Long, M., Acosta, E., Duffy, L.B., Waites, K.B., 2012. Maternal azithromycin therapy for *Ureaplasma* intraamniotic infection delays preterm delivery and reduces fetal lung injury in a primate model. Am. J. Obstet. Gynecol. 207, 475 e1–475 e14.

Grompe, M., Gibbs, R.A., Chamberlain, J.S., Caskey, C.T., 1989. Detection of new mutation disease in man and mouse. Mol. Biol. Med. 6, 511–521.

Guzman, R.E., Evans, M.G., Bove, S., Morenko, B., Kilgore, K., 2003. Mono-iodoacetate-induced histologic changes in subchondral bone and articular cartilage of rat femorotibial joints: an animal model of osteoarthritis. Toxicol. Pathol. 31, 619–624.

Hall, K.E., Holowaychuk, M.K., Sharp, C.R., Reineke, E., 2014. Multicenter prospective evaluation of dogs with trauma. J. Am. Vet. Med. Assoc. 244, 300–308.

Hanly, W.C., Artwohl, J.E., Bennett, B.T., 1995. Review of polyclonal antibody production procedures in mammals and poultry. ILAR J. 37, 93–118.

Hardouin, N., Nagy, A., 2000. Gene-trap-based target site for cre-mediated transgenic insertion. Genesis 26, 245–252.

Hayes, S.H., Radziwon, K.E., Stolzberg, D.J., Salvi, R.J., 2014. Behavioral models of tinnitus and hyperacusis in animals. Front Neurol. 5, 179.

Heffner, H.E., Heffner, R.S., 2007. Hearing ranges of laboratory animals. J. Am. Assoc. Lab. Anim. Sci. 46, 20–22.

Heffner, R.S., Heffner, H.E., 1991. Behavioral hearing range of the chinchilla. Hear. Res. 52, 13–16.

Hill, R.C., Ellison, G.W., Burrows, C.F., Bauer, J.E., Carbia, B., 1996. Ileal cannulation and associated complications in dogs. Lab. Anim. Sci. 46, 77–80.

Hof, P.R., Nimchinsky, E.A., Young, W.G., Morrison, J.H., 2000. Numbers of meynert and layer IVB cells in area V1: a stereologic analysis in young and aged macaque monkeys. J. Comp. Neurol. 420, 113–126.

Hoffman, K.M., Timmel, G.B., Meuli-Simmen, C., Meuli, M., Yingling, C.D., Adzick, N.S., 1996. Experimental fetal neurosurgery: the normal neurology of neonatal lambs and abnormal findings after in utero manipulation. Contemp. Top. Lab. Anim. Sci. 35, 53–56.

Humphreys, I., Wood, R.L., Phillips, C.J., Macey, S., 2013. The costs of traumatic brain injury: a literature review. Clinicoecon. Outcomes Res. 5, 281–287.

Institute, N.T., 2015. Available: http://www.nationaltraumainstitute.org/index.html.

Jean, A., Nyein, M.K., Zheng, J.Q., Moore, D.F., Joannopoulos, J.D., Radovitzky, R., 2014. An animal-to-human scaling law for blast-induced traumatic brain injury risk assessment. Proc. Natl. Acad. Sci. U. S. A. 111, 15310–15315.

Justice, M.J., Noveroske, J.K., Weber, J.S., Zheng, B., Bradley, A., 1999. Mouse ENU mutagenesis. Hum. Mol. Genet. 8, 1955–1963.

Kalangos, A., Sierra, J., Vala, D., Cikirikcioglu, M., Walpoth, B., Orrit, X., Pomar, J., Mestres, C., Albanese, S., Jhurry, D., 2006. Annuloplasty for valve repair with a new biodegradable ring: an experimental study. J. Heart Valve Dis. 15, 783–790.

Kalin, N.H., Shelton, S.E., 1989. Defensive behaviors in infant rhesus monkeys: environmental cues and neurochemical regulation. Science 243, 1718–1721.

Kietzman, H.W., Everson, J.L., Sulik, K.K., Lipinski, R.J., 2014. The teratogenic effects of prenatal ethanol exposure are exacerbated by Sonic Hedgehog or GLI2 haploinsufficiency in the mouse. PLoS One 9, e89448.

Kim, M.K., Choi, H.J., Kwon, I., Pierson 3rd, R.N., Cooper, D.K., Soulillou, J.P., O'Connell, P.J., Vabres, B., Maeda, N., Hara, H., Scobie, L., Gianello, P., Takeuchi, Y., Yamada, K., Hwang, E.S., Kim, S.J., Park, C.G., 2014. The International Xenotransplantation Association consensus statement on conditions for undertaking clinical trials of xenocorneal transplantation. Xenotransplantation 21, 420–430.

Kim, M.K., Lee, J.J., Choi, H.J., Kwon, I., Lee, H., Song, J.S., Kim, M.J., Chung, E.S., Wee, W.R., Park, C.G., Kim, S.J., 2013. Ethical and regulatory guidelines in clinical trials of xenocorneal transplantation in Korea; the Korean xenocorneal transplantation consensus statement. Xenotransplantation 20, 209–218.

King, A.J., 2012. The use of animal models in diabetes research. Br. J. Pharmacol. 166, 877–894.

Lalonde, R., Strazielle, C., 2007. Spontaneous and induced mouse mutations with cerebellar dysfunctions: behavior and neurochemistry. Brain Res. 1140, 51–74.

Lampropoulou-Adamidou, K., Lelovas, P., Karadimas, E.V., Liakou, C., Triantafillopoulos, I.K., Donta, I., Papaioannou, N.A., 2014. Useful animal models for the research of osteoarthritis. Eur. J. Orthop. Surg. Traumatol. 24, 263–271.

Le Floc'h, S., Cloutier, G., Saijo, Y., Finet, G., Yazdani, S.K., Deleaval, F., Rioufol, G., Pettigrew, R.I., Ohayon, J., 2012. A four-criterion selection procedure for atherosclerotic plaque elasticity reconstruction based on in vivo coronary intravascular ultrasound radial strain sequences. Ultrasound Med. Biol. 38, 2084–2097.

Lee, J.J., Kim, D.H., Jang, Y.E., Choi, H.J., Kim, M.K., Wee, W.R., 2014. The attitude toward xenocorneal transplantation in wait-listed subjects for corneal transplantation in Korea. Xenotransplantation 21, 25–34.

Leenaars, M., Hendriksen, C.F., 2005. Critical steps in the production of polyclonal and monoclonal antibodies: evaluation and recommendations. ILAR J. 46, 269–279.

Leiter, E.H., Reifsnyder, P., Driver, J., Kamdar, S., Choisy-Rossi, C., Serreze, D.V., Hara, M., Chervonsky, A., 2007. Unexpected functional consequences of xenogeneic transgene expression in beta-cells of NOD mice. Diabetes Obes. Metab. 9 (Suppl. 2), 14–22.

London, W.T., Martinez, A.J., Houff, S.A., Wallen, W.C., Curfman, B.L., Traub, R.G., Sever, J.L., 1986. Experimental congenital disease with simian cytomegalovirus in rhesus monkeys. Teratology 33, 323–331.

Lucey, J.F., Behrman, R.E., 1963. Thalidomide: effect upon pregnancy in the rhesus monkey. Science 139, 1295–1296.

Malkesman, O., Tucker, L.B., Ozl, J., McCabe, J.T., 2013. Traumatic brain injury – modeling neuropsychiatric symptoms in rodents. Front. Neurol. 4, 157.

Manji, R.A., Menkis, A.H., Ekser, B., Cooper, D.K., 2012. Porcine bioprosthetic heart valves: the next generation. Am. Heart J. 164, 177–185.

McKinney Jr., W.T., Bunney Jr., W.E., 1969. Animal model of depression. I. Review of evidence: implications for research. Arch. Gen. Psychiatry 21, 240–248.

Melvold, R.W., Wang, K., Kohn, H.I., 1997. Histocompatibility gene mutation rates in the mouse: a 25-year review. Immunogenetics 47, 44–54.

Melvold, R.W.K.H.I., 1975. Histocompatibility gene mutation rates: H-2 and non-H-2. Mutat. Res. 27, 415–418.

Mulder, G., Pritchett, K., 2003. The Morris water maze. J. Am. Assoc. Lab. Anim. Sci. 42, 49–50.

Mulder, G., Pritchett, K., 2004. The elevated plus maze. J. Am. Assoc. Lab. Anim. Sci. 43, 39–40.

Myers, P.O., Kalangos, A., 2013. Valve repair using biodegradable ring annuloplasty: from bench to long-term clinical results. Heart Lung Vessel 5, 213–218.

Nagai, K., Yagi, S., Uemoto, S., Tolba, R.H., 2013. Surgical procedures for a rat model of partial orthotopic liver transplantation with hepatic arterial reconstruction. J. Vis. Exp. e4376.

Nappi, F., Spadaccio, C., Castaldo, C., Di Meglio, F., Nurzynska, D., Montagnani, S., Chello, M., Acar, C., 2014. Reinforcement of the pulmonary artery autograft with a polyglactin and polydioxanone mesh in the Ross operation: experimental study in growing lamb. J. Heart Valve Dis. 23, 145–148.

Noveroske, J.K., Weber, J.S., Justice, M.J., 2000. The mutagenic action of N-ethyl-N-nitrosourea in the mouse. Mamm. Genome 11, 478–483.

Peters, A., Verderosa, A., Sethares, C., 2008. The neuroglial population in the primary visual cortex of the aging rhesus monkey. Glia 56, 1151–1161.

Petraglia, A.L., Dashnaw, M.L., Turner, R.C., Bailes, J.E., 2014. Models of mild traumatic brain injury: translation of physiological and anatomic injury. Neurosurgery 75 (Suppl. 4), S34–S49.

Pfannenstiel, T.J., 2014. Noise-induced hearing loss: a military perspective. Curr. Opin. Otolaryngol. Head Neck Surg. 22, 384–387.

Rawlings, C.A., Van Lue, S., King, C., Freeman, L., Damian, R.T., Greenacre, C., Chernosky, A., Mohamed, F.M., Chou, T.M., 2000. Serial laparoscopic biopsies of liver and spleen from Schistosoma-infected baboons (*Papio* spp.). Comp. Med. 50, 551–555.

Roberts, A.J., Cole, M., Koob, G.F., 1996. Intra-amygdala muscimol decreases operant ethanol self-administration in dependent rats. Alcohol. Clin. Exp. Res. 20, 1289–1298.

Roth, A.K., Bogie, R., Jacobs, E., Arts, J.J., Van Rhijn, L.W., 2013. Large animal models in fusionless scoliosis correction research: a literature review. Spine J. 13, 675–688.

Rowe, R.K., Harrison, J.L., Thomas, T.C., Pauly, J.R., Adelson, P.D., Lifshitz, J., 2013. Using anesthetics and analgesics in experimental traumatic brain injury. Lab. Anim. (NY) 42, 286–291.

Russell, W.M.S., Burch, R.L., 1959. The Principles of Humane Experimental Technique. Available Online: http://altweb.jhsph.edu/pubs/books/humane_exp/het-toc, 2015.

Sarr, M., 1988. Pancreas. In: Swindle, M., Adams, R. (Eds.), Experimental Surgery and Physiology: Induced Animal Models of Human Disease. Williams and Wilkins, Baltimore.

Schlager, G., Dickie, M.M., 1971. Natural mutation rates in the house mouse: estimates for five specific loci and dominant mutations. Mutat. Res. 11, 89–96.

Shippenberg, T.S., Koob, G.F., 2002. Recent advances in animal models of drug addiction and alcoholism. In: Davis, K.L., Charney, D., Coyle, J.T., Nemeroff, C. (Eds.), Neuropsychopharmacology: The Fifth Generation of Progress. Lippincott Williams and Wilkins; Philadelphia, PA, pp. 1381–1397.

Sheu, S.Y., Wang, W.L., Fu, Y.T., Lin, S.C., Lei, Y.C., Liao, J.H., Tang, N.Y., Kuo, T.F., Yao, C.H., 2014. The pig as an experimental model for mid-dermal burns research. Burns 40, 1679–1688.

Shofti, R., Zaretzki, A., Cohen, E., Engel, A., Bar-El, Y., 2004. The sheep as a model for coronary artery bypass surgery. Lab. Anim. 38, 149–157.

Skovsø, S., 2014. Modeling type 2 diabetes in rats using high fat diet and streptozotocin. J. Diabetes Investig. 5, 349–358.

Stanley, L., 2014. Molecular and Cellular Toxicology: An Introduction. Wiley, Somerset, NJ, USA.

Stavrakis, A.I., Niska, J.A., Loftin, A.H., Billi, F., Bernthal, N.M., 2013. Understanding infection: a primer on animal models of periprosthetic joint infection. ScientificWorldJournal 2013, 925906.

Streeter, E.M., Rozanski, E.A., Laforcade-Buress, A.D., Freeman, L.M., Rush, J.E., 2009. Evaluation of vehicular trauma in dogs: 239 cases (January−December 2001). J. Am. Vet. Med. Assoc. 235, 405−408.

Struillou, X., Boutigny, H., Soueidan, A., Layrolle, P., 2010. Experimental animal models in periodontology: a review. Open Dent. J. 4, 37−47.

Stucken, E.Z., Hong, R.S., 2014. Noise-induced hearing loss: an occupational medicine perspective. Curr. Opin. Otolaryngol. Head Neck Surg. 22, 388−393.

Suokas, A.K., Sagar, D.R., Mapp, P.I., Chapman, V., Walsh, D.A., 2014. Design, study quality and evidence of analgesic efficacy in studies of drugs in models of OA pain: a systematic review and a meta-analysis. Osteoarthritis Cartilage 22, 1207−1223.

Swartz, D.D., Andreadis, S.T., 2013. Animal models for vascular tissue-engineering. Curr. Opin. Biotechnol. 24, 916−925.

Tanaka, K., Ishizaki, N., Nishimura, A., Yoshimine, M., Kamimura, R., Taira, A., 1993. A new animal model for split liver transplantation using an infrahepatic IVC graft. Surg. Today 23, 609−614.

Tarantal, A.F., Salamat, M.S., Britt, W.J., Luciw, P.A., Hendrickx, A.G., Barry, P.A., 1998. Neuropathogenesis induced by rhesus cytomegalovirus in fetal rhesus monkeys (*Macaca mulatta*). J. Infect. Dis. 177, 446−450.

Tokareva, O., Michalczechen-Lacerda VÍ, A., Rech EÍ, L., Kaplan, D.L., 2013. Recombinant DNA production of spider silk proteins. Microb. Biotechnol. 6, 651−663.

Toscano, M.G., Ganea, D., Gamero, A.M., 2011. Cecal ligation puncture procedure. J. Vis. Exp. 51.

Turner, J.G., Parrish, J.L., Hughes, L.F., Toth, L.A., Caspary, D.M., 2005. Hearing in laboratory animals: strain differences and nonauditory effects of noise. Comp. Med. 55, 12−23.

Valparaiso, A.P., Vicente, D.A., Bograd, B.A., Elster, E.A., Davis, T.A., 2015. Modeling acute traumatic injury. J. Surg. Res. 194, 220−232.

van der Staay, F.J., Arndt, S.S., Nordquist, R.E., 2009. Evaluation of animal models of neurobehavioral disorders. Behav. Brain Funct. 5, 11−33.

Van Meer, P., Raber, J., 2005. Mouse behavioural analysis in systems biology. Biochem. J. 389, 593−610.

Varagona, G., Ellis, L.A., Moore, D., Penney, D., Dusheiko, G.M., 1991. A percutaneous liver biopsy technique in ducks (*Anas platyrhynchos*) experimentally infected with duck hepatitis B virus. Lab. Anim. 25, 254−257.

Venter, N.G., Monte-Alto-Costa, A., Marques, R.G., 2014. A new model for the standardization of experimental burn wounds. Burns 41, 542−547.

von der Behrens, W., 2014. Animal models of subjective tinnitus. Neural Plast. 2014, 741452.

Voss, W.R., 1970. Primate liver and spleen biopsy procedures. Lab. Anim. Care 20, 995−997.

Wang, Y., Zhao, S., Bai, L., Fan, J., Liu, E., 2013. Expression systems and species used for transgenic animal bioreactors. Biomed. Res. Int. 2013, 580463.

Wei, Q., Dong, Z., 2012. Mouse model of ischemic acute kidney injury: technical notes and tricks. Am. J. Physiol. Ren. Physiol. 303, F1487−F1494.

Weiss, F., 2010. Advances in animal models of relapse for addiction research. In: Kuhn, C., Koob, G. (Eds.), Advances in the Neuroscience of Addiction. CRC Press, Boca Raton, FL.

Williams, D., 2003. Sows' ears, silk purses and goats' milk: new production methods and medical applications for silk. Med. Device Technol. 14, 9−11.

Wingfield, W.E., Tumbleson, M.E., Hicklin, K.W., Mather, E.C., 1974. An exteriorized cranial vena caval catheter for serial blood sample collection from miniature swine. Lab. Anim. Sci. 24, 359–361.

Wronski, T.J., Lowry, P.L., Walsh, C.C., Ignaszewski, L.A., 1985. Skeletal alterations in ovariectomized rats. Calcif Tissue Int. 37, 324–328.

Xia, B., Di, C., Zhang, J., Hu, S., Jin, H., Tong, P., 2014. Osteoarthritis pathogenesis: a review of molecular mechanisms. Calcif Tissue Int. 95, 495–505.

Zeyland, J., Lipinski, D., Slomski, R., 2015. The current state of xenotransplantation. J. Appl. Genet. 56, 211–218.

Commonly Used Animal Models

7

D.L. Hickman[1], J. Johnson[2], T.H. Vemulapalli[3], J.R. Crisler[1], R. Shepherd[4]

Indiana University, Indianapolis, IN, United States[1]; The University of the West Indies, Trinidad, West Indies[2]; Purdue University, West Lafayette, IN, United States[3]; Indiana University, Bloomington, IN, United States[4]

CHAPTER OUTLINE

Principles of Animal Research for Graduate and Undergraduate Students.
Copyright © 2017 Elsevier Inc. All rights reserved.

7.1 THE MOUSE

7.1.1 INTRODUCTION

The mouse is a small mammal that belongs to the order Rodentia (Fig. 7.1). The house mouse of North America and Europe, *Mus musculus,* is the species most commonly used for biomedical research. It is likely that the mouse originated in Eurasia and utilized its commensal relationship with humans to spread through to other continents as humans explored and colonized. Mouse fanciers around the turn of the 20th century are the source of the majority of the laboratory mice that are in use today. A summary of the overarching categories of mouse models that are available is presented in Table 7.1.

7.1.2 USES IN RESEARCH

Mice and rats make up approximately 95% of all laboratory animals, with mice the most commonly used animal in biomedical research. Mice are a commonly selected animal model for a variety of reasons, including small size (facilitating housing and maintenance); short reproductive cycle and lifespan; generally mild-tempered and docile; wealth of information regarding their anatomy, genetics, biology, and physiology; and the possibility for breeding genetically manipulated mice and mice that have spontaneous mutations.

Mice have been used as research subjects for studies ranging from biology to psychology to engineering. They are used to model human diseases for the purpose of finding treatments or cures. Some of the diseases they model include: hypertension, diabetes, cataracts, obesity, seizures, respiratory problems, deafness, Parkinson's disease, Alzheimer's disease, various cancers, cystic fibrosis, and acquired immunodeficiency syndrome (AIDS), heart disease, muscular dystrophy, and spinal cord injuries. Mice are also used in behavioral, sensory, aging, nutrition, and genetic studies. This list is in no way complete as geneticists, biologists, and other scientists are rapidly finding new uses for the domestic mouse in research.

7.1.3 NORMATIVE BIOLOGY

Mice are mammals and their organ systems are very similar to organ systems in humans in terms of shape, structure, and physiology. Basic physiologic data are presented in Table 7.2.

FIGURE 7.1

Laboratory Mouse.

Photo used with permission of American Association for Laboratory Animal Science.

Table 7.1 Summary of the Various Categories of Mouse Types That Are Available for Use in Research

Model	Generation	Uses	Examples
Inbred strains	20 or more consecutive generations of sister × brother or parent × offspring matings	Studies that require genetically identical animals	BALB/c, C3H, C57BL/6, CBA, DBA/2, C57BL/10, AKR, A, 129, SJL
Outbred stocks	Deliberate mating of unrelated animals	Studies that require outbred vigor	Swiss Webster, CD-1, ICR
Spontaneous mutant	Strains that have been bred to conserve phenotypical characteristics that were due to spontaneous genetic mutations	Studies of disease processes associated with the spontaneous mutation	Athymic nude, nonobese diabetic (NOD)
Genetically engineered mice/ "Knock-in"/ "Knock-out"	Mice where genes have been turned on or off	Studies that are seeking to identify the effects of specific genes	
Transgenic	Mice where a gene from an unrelated species has been inserted into the genome	Studies that require a mouse model of human disease and toxicology	

Mice have very long loops of Henle in the kidneys, thus allowing for maximal concentration of their urine. As a result, urine output in mice usually consists of only a drop or two of highly concentrated urine at a time. They also excrete large amounts of protein in their urine with sexually mature male mice excreting the largest levels of protein possibly as pheromones.

Mice have only two types of teeth, incisors and molars. The incisors are open-rooted and erupt (i.e., grow) continuously throughout their lives. This predisposes mice to malocclusion if not given feeds or objects such as nylon bones to help wear down the teeth during mastication. The molars are rooted and, thus, do not continuously erupt.

The stomach has two compartments with the proximal portion completely keratinized and the distal portion entirely glandular. Their intestines are simple, but the rectum is very short (1−2 mm) and hence is prone to prolapse, especially if the animal has colitis. The gastrointestinal flora consists of more than 100 species of bacteria that form a complex ecosystem that aids digestion and health of the mouse.

Table 7.2 Basic Mouse Biological Data

Adult weight	
Male	20–40 g
Female	25–40 g
Life span	1–3 years
Body temperature	97.7°F–100.4°F (36.5°C–38.0°C)
Heart rate	310–840/min
Respiration rate	163 breaths/min
Water consumption	5–8 mL/day
Food consumption	3–5 g/day

Mice have no sweat glands but have a relatively large surface area per gram of body weight. This results in dramatic changes in physiology and behavior in response to fluctuations in ambient temperature. When too cold, mice will respond by nonshivering thermogenesis (i.e., metabolism of brown adipose tissue). In addition to the lack of sweat glands, they cannot pant or produce large amounts of saliva to aid in cooling their body temperature. Therefore, when exposed to very hot situations, mice increase the blood flow to their ears to maximize heat loss; and in the wild, they move into their burrows which are at cooler temperatures. The thermoneutral zone, the range of ambient temperatures at which the mouse does not have to perform regulatory changes in metabolic heat production or evaporative heat loss to maintain its core temperature, is about 85.3°F–86.9°F (29.6°C–30.5°C).

7.1.4 REPRODUCTIVE PHYSIOLOGY

The female reproductive system is comprised of paired ovaries and oviducts, uterus, cervix, vagina, clitoris, and paired clitoral glands. Pregnant female mice have hemochorial placentation, similar to humans (i.e., maternal blood is in direct contact with the chorion, the outermost layer of the fetal placental membranes). The female mouse also has five pairs of mammary glands. The male reproductive system consists of paired testes, penis, and associated sexual ducts and glands. The inguinal canals are open in the male mouse, and the testes can retract easily into the abdominal cavity. Both sexes have well-developed preputial glands, which can become infected. Males have a number of accessory sex glands, including large seminal vesicles, coagulating glands, and a prostate. Secretions from these glands make up a large part of the mouse's ejaculate. When mice ejaculate, the semen forms a coagulum or copulatory plug.

Mice breed continuously throughout the year unless conditions are very unfavorable to them (e.g., lack of food). Their reproductive potential can be affected by a number of external influences such as noise, diet, light cycles, population density, or cage environment. Genotype also can affect reproductive performance as it is common knowledge that some inbred strains of mice are poor breeders, and if pups are born they may receive poor maternal care. Additional reproductive physiologic data are presented in Table 7.3.

Mice can be bred using a one-on-one system (one male to one female; monogamous) or in a harem mating system (polygamous mating). In a monogamous system, the male and female are always left together, but at weaning the pups are removed from the cage. This system allows for maximal use of the postpartum estrus and the maximum number of litters for the females involved, and facilitates record-keeping and monitoring of specific breeders in the colony. In a harem mating system, multiple males are placed with multiple females, usually at a ratio of one male to two to six females. Usually, females are removed to separate cages just before parturition and the postpartum estrus is underutilized.

Mouse pups are born hairless, blind, and deaf and require extensive parental care that is provided mainly by the mother. Due to the ruddy coloration of the skin of the hairless pups, they are also known as "pinkies." While mouse pups can increase their body temperature through the metabolism of brown fat stores, they are unable to adequately conserve body heat until they develop an adequate fur coat. Thus, inclusion of nesting materials in the cage is highly recommended as huddling inside the nest can provide much needed warmth and safeguard against temperature-associated neonatal losses.

The most reliable method for determining the sex of a mouse is by measuring the length of the anogenital distance, i.e., the distance from the anus to the genitalia. This distance can be measured with a ruler or animals assessed side by side with the rear ends held up by their tails. The anogenital distance is longer in males than females. In sexually mature animals one can also determine the sex of mice by the presence or absence of testicles in a testicular sac.

7.1.5 MOUSE BEHAVIOR

Mice are social creatures and can be group housed easily. Their main method of communication is via pheromones. They use these olfactory cues to establish a

Table 7.3 Basic Mouse Reproductive Reference Values

Puberty	
Male	4–7 weeks
Female	4–8 weeks
Estrus cycle	4–5 days
Gestation	19–21 days
Litter size	9–12 pups
Weaning	21 days
Breeding duration	7–9 months

pecking order (i.e., a hierarchical system of social organization). These chemicals are so important that when cage environments are changed, such as by simple cleaning or with bedding changes, a bout of fighting may occur until scent marking of the cage is completed as a way to reestablish the pecking order and social organization in that cage.

Pheromones also play a vital role in reproduction of these animals. This is demonstrated by the Whitten effect and the Bruce effect. The Whitten effect occurs when a group of female mice that are not cycling are exposed to male urine, which contains a large quantity of pheromones. The females will all resume cycling as a group soon after the introduction of the male. In contrast, the Bruce effect is characterized by abortion of litters when pregnant females are exposed to the urine of a strange male.

As with most rodents, mice are nocturnal animals exhibiting peak levels of activity at night. Because mice are a prey species, they display thigmotactic behavior or wall hugging. They avoid open spaces where they might be easily caught by predators. Despite this, mice are very curious about any new objects in their territory and will often examine them at length.

Mice not only have poorly developed eyesight but they are also color blind. A number of inbred strains (e.g., FVB/N and C3H/He) are functionally blind by weaning. They rely on their very sensitive hearing to escape detection and on their sense of smell and taste to detect food (and possibly avoid poisons). Mice can hear over a range of frequencies between 0.5 and 120 kHz; however, normal mice are most sensitive to frequencies of 12−24 kHz. It is important to note here that some inbred strains of mice, e.g., C57BL/6, suffer significant hearing loss before 1 year of age.

Mice can climb, swim, and jump (up to a foot), though they normally prefer to avoid swimming, if possible. Under certain conditions they display stereotypies, which are obsessive−compulsive behaviors. The behaviors may be strain-related, environment-related, or study-related and include wire gnawing, circling, jumping, and aggression. The use of environmental enrichment items such as cardboard tubes or other structures offer the animal an area for retreat from cage mates and add complexity to the environment.

Aggression is another important behavior that commonly occurs in group-housed male mice. It can also occur in group-housed females and mixed-sex cages. Indications that there is an aggressive animal in the cage include bite wounds on the tail, rump, ears, and shoulders of mice (Fig. 7.2). The wounds can be so severe as to cause significant blood loss and abscess formation at bite sites. Aggression has been shown to be influenced by strain, age, and prior encounters. In terms of strains, the more aggressive strains are BALB/c, C57BL/10, C57BL/6, DBA/2, and outbred Swiss. Methods to prevent or reduce aggression include use of properly designed enrichment devices; provision of adequate space and shelter for each animal; grouping of mice before they reach puberty; use of docile strains; and removal of dominant animals as soon as possible.

FIGURE 7.2

Lesions caused by aggression in mice.

Photograph provided by Jenelle Johnson.

A common manifestation of social organization in group-housed mice is barbering, a behavior in which a dominant mouse will trim, by chewing, the hair or whiskers of other mice in the cage. Barbering is also sometimes referred to as the Dalila effect. It is usually instigated by a dominant male or female mouse. The dominant animals retain their whiskers and full hair coats, while their cage mates have "shaved faces and bodies" (Fig. 7.3) (Garner et al., 2004). Although barbering does not generally result in any physical harm to the animal, removing the dominant mouse (the nonbarbered one) from the cage is a good approach to control.

7.1.6 HOUSING AND HANDLING

General types of housing mice in a laboratory setting include: **conventional**, **specific pathogen free** (SPF), and **germ free**. In conventional housing, no attempt is made to keep out adventitious microbial and parasitic organisms. Mice housed in this manner can be found in open-topped cages. Room air, along with any airborne contaminants, is allowed to freely circulate into the mouse's cage. In addition, the food and water are not sterilized, though it should be noted that microbial contaminants may enter into the mouse population in this way.

SPF mice are raised in barrier conditions to ensure that they remain free of a specific list of pathogens. Care is taken to ensure that adventitious microbes and parasites are excluded from the animals. SPF mice are typically raised in specialized

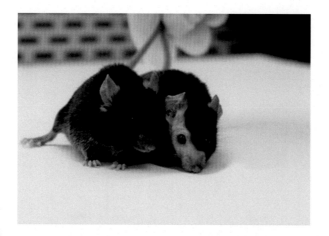

FIGURE 7.3

An example of barbering of one mouse (right) by the other mouse (left).

Photograph provided by Jenelle Johnson.

caging such as microisolator cages. These cages contain a 0.22 μm filter top that aids in the exclusion of microbes and parasites. Individually ventilated caging systems include a rack of microisolator cages, each of which receives a filtered air supply (Fig. 7.4). Under SPF conditions, everything that comes into contact with the animal should be sterilized or disinfected. This includes, but is not limited to, the water, food, bedding, and caging. Special care must be taken by anyone handling the mice, including researchers, to ensure that handling and experimental procedures do not introduce potential pathogens into the colony. Thus, all handling and procedures done on SPF mice are often performed under HEPA-filtered air conditions, such as within a biosafety cabinet. Placing the mouse in an unfiltered environment ("room" air), even for a moment, is enough to potentially colonize the mouse with a whole host of adventitious microorganisms, thus destroying its SPF status. Once a contaminated mouse is placed back into the colony, the entire colony is at risk for infection.

Germ-free, or axenic, mice are raised to contain no microbes or parasites whatsoever. Raising germ-free mice requires strict barrier maintenance. Usually, this requires the use of flexible film isolators which provide HEPA-filtered air to the mice within the isolator. Additionally, any materials (e.g., food, water, and bedding) must be sterilized or thoroughly disinfected prior to being moved into the isolator unit so as not to contaminate the living space of the germ-free rodents with adventitious microorganisms. All procedures performed within the flexible film isolator must utilize strict aseptic technique for the same reason.

In addition to the microbiological environment of the animal's housing systems, mice need to be housed at specific environmental parameters otherwise they may experience stress. The *Guide for the Care and Use of Laboratory Animals,*

FIGURE 7.4

Individually ventilated caging system for mice.

Photo provided by Kay Stewart.

8th edition (National Research Council, 2011) is an internationally accepted document that outlines and discusses globally accepted environmental parameters for housing different species of animals including the mouse. Table 7.4 outlines the specific environmental requirements listed in this document for housing mice.

Mice are omnivorous and coprophagic with at least one-third of their diet being the consumption of their feces. In the laboratory setting mice are fed a clean, wholesome, and nutritious pelleted rodent diet ad lib. There are many commercially formulated diets for the various stages of life and for animals with specific induced diseases such as diabetes mellitus or hypertension. These diets may be available as autoclavable or irradiated forms to prevent transmission of disease via contaminated feed. There are also a variety of "pet" treats available for mice. However, the treats should not make up more than 5–10% of the daily diet.

Mice should be provided with a continuous supply of water daily. If the animals do not get enough water daily, their food consumption will decrease. The animals will also look scruffy and unhealthy. Mice can be provided with water from water bottles or pouches, automatic watering systems with nipples, or water-based gel packs.

Table 7.4 Environmental Parameters for Mice

Dry-bulb macroenvironmental temperature	20–26°C (68–79°F)
Relative humidity	30–70%
Air changes per hour in the animal rooms	10–15 ACH
Light	130–325 lux 1 m (3.3 ft.) above the floor
Floor space requirements	
Weight: <10 g	6 in.^{2} (38.7 cm^2)
Up to 15 g	8 in.^{2} (51.6 cm^2)
Up to 25 g	12 in.^{2} (77.4 cm^2)
>25 g	≥15 in.^{2} (≥96.7 cm^2)
Female and litter	51 in.^{2} (330 cm^2)
Cage height requirements	5 in. (12.7 cm)

7.1.7 DISEASES

Some general signs of ill health include: weight loss, depression or lethargy, anorexia, obesity, diarrhea, scruffiness or ruffled coats, abnormal breathing, sneezing, weakness, dehydration, enlarged abdomen, discolorations (e.g., yellow for jaundiced animals or very pale for anemic animals), masses or swellings, and abnormal posture or gait. Body condition scoring is an objective measure to truly assess how fat or thin the animal is and can be used for accurate determination of endpoints in studies where animals are expected to lose or gain weight (Fig. 7.5) (Ullman-Cullere and Foltz, 1999). Some of the more commonly found diseases of mice are presented in Table 7.5. Noninfectious disorders are presented in Table 7.6.

Based on their genetic and physiologic makeup, mice can be either immunocompetent or immunodeficient. **Immunocompetent** means that the mouse has a normal functioning immune system and can stage an immune response to any insult or injury. In contrast, **immunodeficient** means that some component or components of the mouse's immune system is not working or functioning normally, and so they cannot stage an adequate immune response and are more susceptible to infectious disease. Immunosuppressed mice are mice that have a complete immune system but because of a drug or chemical or disease state, the immunological response is attenuated.

7.2 THE RAT
7.2.1 INTRODUCTION

Rats and humans have a long history of coexistence. The origins of the laboratory rat, also known as the Norway rat, stretch back centuries to the areas of modern day China and Mongolia (Burt, 2006; Song et al., 2014). The dispersal of the

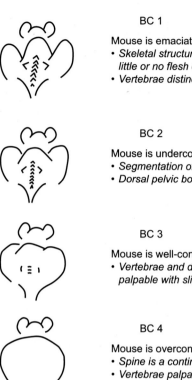

BC 1

Mouse is emaciated.
- *Skeletal structure extremely prominent; little or no flesh cover.*
- *Vertebrae distinctly segmented.*

BC 2

Mouse is underconditioned.
- *Segmentation of vertebral column evident.*
- *Dorsal pelvic bones are readily palpable.*

BC 3

Mouse is well-conditioned.
- *Vertebrae and dorsal pelvis not prominent; palpable with slight pressure.*

BC 4

Mouse is overconditioned.
- *Spine is a continuous column.*
- *Vertebrae palpable only with firm pressure.*

BC 5

Mouse is obese.
- *Mouse is smooth and bulky.*
- *Bone structure disappears under flesh and subcutaneous fat.*

A "+" or a "ˌ" can be added to the body condition score if additional increments are necessary (i.e. …2+, 2, 2–…)

FIGURE 7.5

Mouse body condition (BC) scoring chart (Ullman-Cullere and Foltz, 1999).

Norway rat has occurred across the centuries and its natural habitat stretches from the Mediterranean across Southeast Asia and down into Australia and New Guinea (Song et al., 2014). Unfortunately, most people associate rats with disease and destruction. Throughout history, outbreaks of bubonic plague, typhus, and hantaviruses have had an unwitting accomplice in the rat (Zinsser, 1935; Benedictow, 2004; Firth et al., 2014). Over the centuries, rats have also been used for food

Table 7.5 Infectious Diseases of Mice

Pathogen	Common Names of Diseases	Fatal/Self-Limiting	Effects on Research
Parvoviruses	Minute virus of mice (MVM) Mouse parvovirus (MPV)	Self-limiting Asymptomatic	Can distort biological responses that depend on cell proliferation, primarily on immune cells such as cytotoxic T cells.
Lymphocytic Choriomeningitis virus (LCMV)		IC—self-limiting and asymptomatic In others it can lead to clinical signs and death	Death of animals on research Affects immunological function Zoonotic
Reoviruses	Reovirus 1, 2, or 3	Disease and mortality in suckling mice	High mortality in young mice May be oncolytic Interference with organ systems
	Rotavirus (Epizootic diarrhea of Infant mice)	Infection primarily in young mice but self-limiting	Clinical illness leads to diarrhea and retarded growth.
Coronaviruses	Mouse hepatitis virus (Mouse coronaviruses)	IC—asymptomatic IDS and suckling mice—clinical disease and death	Numerous effects including affecting immunological responses, contaminate transplantable neoplasms, altering tissue enzymes, and causing clinical disease.
Mycoplasmosis	*Mycoplasma pulmonis* *Mycoplasma arthritidis* *Mycoplasma neurolyticum* *Mycoplasma collis*	Rhinitis Otitis media Chronic pneumonia Infertility Can be fatal	Cause clinical disease Compromise experiments involving respiratory tract Alter immunological responses Contamination of cell lines and tumors
Cilia-associated respiratory (CAR) bacillus	*Bacillus sp.*	Respiratory disease Clinical disease—rare	

Continued

Table 7.5 Infectious Diseases of Mice—cont'd

Pathogen	Common Names of Diseases	Fatal/Self-Limiting	Effects on Research
Helicobacteriosis	*Helicobacter hepaticus Helicobacter bilis, Helicobacter muridarum, Helicobacter rodentium, Helicobacter typhlonius*	IC— asymptomatic IDS— inflammatory bowel disease, Diarrhea, rectal prolapse	Affects liver enzymes in some strains of mice Morbidity and mortality in mice Alter immune function
Mycobacteriosis	*Mycobacteria avium-intracellulare* *Mycobacteria leprae-murium*	Granulomatous pneumonia and microgranulomas in other major organs	Morbidity
Pneumocystosis	*Pneumocystis carinii*	IC—asymptomatic IDS— severe disease; pneumonia May be fatal	Pneumonia in IDS mice; may be fatal
Nematodes (Mouse pinworms)	*Syphacia obvelata Aspicularis tetraptera*	Asymptomatic with gastrointestinal lesions such as enteritis and fecal impaction	Unthriftiness in mice Alter the immune responses in animal
Acariasis (Mite infestation)	*Myobia musculi Myocoptes musculinus Radfordia affinis*	Skin lesions, excessive pruritus (scratching), hair loss, dermatitis	Disrupt immunological responses Severe clinical disease

IC, *immune competent;* IDS, *immune compromised.*

Table 7.6 Noninfectious Disorders of Mice

Disorder	Cause	Effects on Research
Hypothermia	Room environment too cold	Can distort biological responses of the mouse due to cold stress
Hyperthermia	Room environment too hot	Death of animals
Ringtail	Relative humidity of less than 30%; causes circular constrictions of the tail	Can lead to loss of tail
Malocclusion	Genetic; poor alignment of the incisors leads to overgrown teeth	Inability to eat

(e.g., in Imperial China), companionship, and sport (Gorn and Goldstein, 2004; Hopkins et al., 2004; Burt, 2006). Ratting, a vicious blood sport where people laid bets on the dog that could kill the most rats in a given period of time was especially popular in both the Victorian England and American Underworld (Thomas and Mayhew, 1998; Gorn and Goldstein, 2004). At the turn of the 20th century, breeding rats as a hobby or for companionship (i.e., "fancy rats") was recognized by the addition of "rat" to both the name and mission of National Mouse Club in the United Kingdom (American Fancy Rat and Mouse Association, 2014). However, as interest in pet rats waned over the following years, the club reorganized and dropped "rat" from its name. A similar club, the American Fancy Rat and Mouse Association, was founded in the United States in 1983 (American Fancy Rat and Mouse Association, 2014).

7.2.2 USES IN RESEARCH

The first recorded use of rats as research subjects occurred in 1828 (Hedrich, 2000), and the first known rat breeding experiments occurred in the late 1800s (Lindsey and Baker, 2006). The first major effort to perform research in the United States using laboratory rats occurred at the Wistar Institute of Philadelphia, the oldest independent research institute in the United States, in 1894 (Lindsey and Baker, 2006).

Rattus norvegicus constitutes one of the most commonly used laboratory species (Fig. 7.6), second only to the laboratory mouse. Because rats and mice are not included under the Animal Welfare Act Regulations, the precise number of these species used per year within the United States is unavailable. However, examining the data collected within the European Union can give some indication of their use relative to other common laboratory animal species. In 2011, rats accounted for just fewer than 14% (1.6 million) of the total animals (11.5 million) used in research within the European Union (European Commission, 2013). This contrasts to mice, which constituted 61% (6.9 million) of the total animals used within the European Union (European Commission, 2013).

Rats possess a number of qualities which make them a highly suitable and much preferred animal model. Like mice, these traits include relatively small size; known genetic background; short generation time; similarities to human disease conditions; and known microbial status. Their tractable nature makes them easier to handle in a laboratory setting than many other rodents. Rats rarely bite their handlers unless extremely stressed or in pain.

Rats have been used as animal models in numerous areas of research from space exploration to answering more basic scientific questions regarding nutrition, genetics, immunology, neurology, infectious disease, metabolic disease, and behavior. Perhaps their largest use is in drug discovery, efficacy, and toxicity studies. In the United States, the approval of any new drug for use in humans or animals usually necessitates that toxicity testing be done in at least one small animal species (e.g., rodents) and one large animal or target species (e.g., dog, nonhuman primate).

FIGURE 7.6

Two juvenile Lobund-Wistar laboratory rats.

Photo provided by Kay Stewart.

7.2.3 NORMATIVE BIOLOGY

There are known physiologic differences between the numerous outbred stocks and inbred strains of rats. The Rat Genome Database (RGD) is an extensive, free resource filled with information regarding the different phenotypes, models, and genomic tools used in rat research (Laulederkind et al., 2013). Vendors of commercially available rat stocks and strains are often good resources for normal physiologic data of these strains. Many provide stock- and strain-specific data directly on their websites such as growth curves, complete blood count and serum biochemistry panels, and spontaneous lesions seen on histopathology. A summary table of normal physiologic references is in Table 7.7.

7.2.4 REPRODUCTIVE BIOLOGY

Sexual dimorphism exists between male and female rats. Sexing of adult rats is most easily done by examining the perineal area of the rat and identifying the external reproductive structures such as the penis, testes, or vagina. In addition, male rats are typically larger and weigh significantly more than their age- and strain-

Table 7.7 Basic Rat Biological Data

Adult weight	
Male	260–1000 g
Female	225–500 g
Life span	1–3 years
Body temperature	97.7°F–100.4°F (36.5°C–38.0°C)
Heart rate	310–493/min
Respiration rate	145 breaths/min
Water consumption	15–20 mL/day
Food consumption	22–33 g/day

matched female counterparts. Sexing of rat pups is most easily performed by examining the distance between the anus and genital opening in the pup. Males have a greater anogenital distance than females.

Male rats possess paired testicles that descend from the abdomen into the scrotal sac at approximately 15 days of age (Russell, 1992). Due to the lack of closure of the inguinal rings, the testes may be retracted into the abdominal cavity even as an adult. The male rat also possesses a number of accessory sex glands. A four-lobed prostate is present along with four other paired glands, including: the seminal vesicles, coagulating glands, ampullary glands, and bulbourethral glands (noted in some texts by the older name, Cowper's gland) (Popesko et al., 1992). Due to the unusual bihorned shape of the closely associated coagulating gland and seminal vesicles, individuals unfamiliar with rodent male anatomy may initially mistake these structures for the female uterus. However, the apices of these glands are freely mobile and easily exteriorized from the abdominal cavity unlike the uterus, which is attached to the dorsal body wall bilaterally via paired ovaries and their respective ovarian ligaments.

The reproductive anatomy of the female rat contains some distinct features. The uterus of the female rat is classified as a duplex uterus, because the vagina is separated from the uterus by two individual cervices with each cervix leading to a separate uterine horn (Popesko et al., 1992). The placentation of the pregnant rat is hemotrichorial (three layers) rather than the hemomonochorial (single layer) placentation present in humans (Wooding and Burton, 2008). The rest of the reproductive anatomy (e.g., ovary, oviduct) is structurally and functionally similar to other mammals. A summary of basic reproductive physiology is presented in Table 7.8.

Rat pups are born hairless, blind, and deaf and require extensive parental care that is provided mainly by the mother. As with mice, the skin of the hairless rat pups has a pink coloration, thus they are also referred to as "pinkies." The inclusion of nesting materials in the cage is recommended to assist the rat pups with thermal regulation until they have a full hair coat (Whishaw and Kolb, 2005).

Table 7.8 Basic Rat Reproductive Reference Values

Puberty	
Male	45 days
Female	33–42 days
Estrus cycle	4–5 days
Gestation	21–23 days
Litter size	9–12 pups
Weaning	21–28 days
Breeding duration	7–9 months

7.2.5 NORMAL BEHAVIOR

Like other rodents, *Rattus norvegicus* is a nocturnal species with the highest level of activity occurring during the dark phase. Behaviors exhibited by rats include grooming, nesting, eating, and other social behaviors. Nesting behavior serves several purposes among rats and mice (Gaskill et al., 2012, 2013a,b,c). Nests allow for better thermoregulatory control within a given environment as well as protection against predation (Gaskill et al., 2013c). Recent work in mice suggests that energy not diverted to thermoregulation can be shunted to other activities as seen via improved feed conversion and breeding performance (Gaskill et al., 2013c). However, anecdotal evidence suggests that nest building in rats is largely a learned behavior, and it appears that there is a developmental period in young rats whereupon if exposed to nesting materials during this time they will begin using the materials to build at least rudimentary nests (Gaskill, 2014). At a minimum, rats benefit from having a structural shelter or nest box into which they may rest away from prying eyes (Fig. 7.7).

Rats emit alarm vocalizations during times of distress. However, these negatively associated vocalizations typically register in the ultrasonic wavelengths (approximately 22 kHz), well outside of the human hearing range (Burman et al., 2007; Parsana et al., 2012). Rodents may also emit high-pitched *audible* vocalizations when extremely alarmed, distressed, or in pain (Jourdan et al., 1995; Han et al., 2005).

As discussed in Chapter 5, rats exhibiting abnormal behaviors and stereotypies can create variables in the research findings and should not be used in a study unless abnormal behavior is the object of the study subjects (Baenninger, 1967; Callard et al., 2000; Garner and Mason, 2002; Cabib, 2005; Ibi et al., 2008). Examples of stereotypies seen in rats include: bar-gnawing, pawing behavior, repetitive circling, and backflipping. It is critical that rats should be provided some form of environmental enrichment to stimulate positive species-typical behaviors.

7.2.6 HOUSING

The housing of rats in a laboratory setting is similar to that described previously for mice: conventional, SPF, and germ free. As rats are social animals, at minimum, they

FIGURE 7.7

A multilevel cage with an intracage shelter. This style of caging provides opportunities for exercise for the rats.

Photo provided by Melissa Swan.

should be housed in pairs whenever possible. There is a preponderance of evidence that shows the differences in affiliative versus aggressive behavior, biochemical changes, and changes in learning between rats raised and housed in social isolation versus those housed in social groups (Baenninger, 1967; Einon and Morgan, 1977; Robbins et al., 1996).

Enrichment items, such as a hut, nesting box, or similar type of shelter may be included in the cage to provide a visual barrier between the rat and the rest of the animal room. Evidence suggests that rats prefer shelters made from opaque plastic (Patterson-Kane, 2003). Rats also spend a lot of time in the wild chewing either for eating or for manipulating objects for nest building. Providing objects made of safe materials in the cage allows the rats to exhibit this natural behavior and encourage the normal wear of the rat's incisors, minimizing the incidence of malocclusion of the teeth.

Rodents can benefit from frequent gentle handling by the researcher and animal care staff. This concept is also known as "gentling" and has been demonstrated to reduce the stress experienced by rats during experimental handling and procedures (Hirsjarvi et al., 1990; van Bergeijk et al., 1990). Another positive interaction between humans and rats is found in the "tickling" of rodents. Based on ultrasonic vocalization data, rodents find tickling a pleasurable experience (Burgdorf and Panksepp, 2001; Panksepp, 2007; Hori et al., 2014). Tickling may also decrease the stress response seen in rodents after experimental manipulations like intraperitoneal injections (Cloutier et al., 2014).

7.2.7 DISEASES

As for mice, some general signs of ill health would include: weight loss, depression or lethargy, anorexia, obesity, diarrhea, scruffiness or ruffled coats, abnormal breathing, sneezing, weakness, dehydration, enlarged abdomen, discolorations (e.g., yellow for jaundiced animals or very pale for anemic animals), masses or swellings, and abnormal posture or gait. When assessing animals, a body condition score can be used as an objective measure or scale to truly assess how fat or thin the animal is; and it allows for the accurate determination of endpoints in studies where animals are expected to lose or gain weight (Hickman and Swan, 2010) (Fig. 7.8). Some of the more commonly found diseases of rats are presented in Table 7.9.

7.3 THE RABBIT

7.3.1 INTRODUCTION

The ancestral home of the European rabbit (*Oryctolagus cuniculus*) is the Iberian Peninsula (Hardy et al., 1995). The earliest archeological evidence of the coexistence of humans and rabbits can be found in excavation sites dated at approximately 120,000 years BCE in Nice, France (Dickenson, 2013). In antiquity, Romans used rabbits as a food source and are thought to be responsible for their dispersal throughout Europe, although there is no evidence that they attempted to actually domesticate them (Dickenson, 2013). Domestication and selective breeding is thought to have begun in France in the Middle Ages where monks began to breed rabbits in their monasteries (Dickenson, 2013). The rabbits were kept confined in enclosures called "clapiers"(Dickenson, 2013). They were kept largely as a source of food for the monks especially since 600 CE when Pope Gregory I officially classified them as "fish" and thus eligible to being eaten during Lent. However, rabbit wool production soon became a welcome by-product of these domestication efforts.

7.3.2 USES IN RESEARCH

European rabbits have been used in research since the middle of the 19th century. Early work with the species was concentrated on the comparative anatomy of the rabbit with other species, such as the frog, and the unique features of the rabbit's heart and circulatory system (Champneys, 1874; Roy, 1879; Smith, 1891). Louis Pasteur used rabbits in a series of experiments that led to the development of the world's first rabies vaccine (Rappuoli, 2014).

While there are numerous so-called "fancy" breeds of rabbits available in the pet trade, the list of breeds routinely used in research is much shorter. The New Zealand White (NZW) rabbit is the most frequent breed of used in research (Fig. 7.9). The California and Dutch-belted rabbit breeds are also occasionally used. Researchers have developed genetically inbred rabbit strains for particular research applications. For example, the Watanabe heritable hyperlipidemic (WHHL) and the myocardial

BC 1
Rat is emaciated
- Segmentation of vertebral column prominent if not visible.
- Little or no flesh cover over dorsal pelvis. Pins prominent if not visible.
- Segmentation of caudal vertebrae prominent.

BC 2
Rat is under conditioned
- Segmentation of vertebral column prominent.
- Thin flesh cover over dorsal pelvis, little subcutaneous fat. Pins easily palpable.
- Thin flesh cover over caudal vertebrae, segmentation palpable with slight pressure.

BC 3
Rat is well-conditioned
- Segmentation of vertebral column easily palpable.
- Moderate subcutaneous fat store over pelvis. Pins easily palpable with slight pressure.
- Moderate fat store around tail base, caudal vertebrae may be palpable but not segmented.

BC 4
Rat is overconditioned
- Segmentation of vertebral column palpable with slight pressure.
- Thick subcutaneous fat store over dorsal pelvis. Pins of pelvis palpable with firm pressure.
- Thick fat store over tail base, caudal vertebrae not palpable.

BC 5
Rat is obese
- Segmentation of vertebral column palpable with firm pressure; may be a continuous column.
- Thick subcutaneous fat store over dorsal pelvis. Pins of pelvis not palpable with firm pressure.
- Thick fat store over tail base, caudal vertebrae not palpable.

FIGURE 7.8

Rat body condition (BC) scoring chart (Hickman and Swan, 2010).

infarction-prone WHHL rabbit (WHHLMI), both developed by researchers in Japan, are used to explore diseases associated with dyslipidemia such as atherosclerosis (Shiomi et al., 2003; Shiomi and Ito, 2009).

Rabbits have been used as a model of human pregnancy and for the production of polyclonal antibodies for use in immunology research (Hanly et al., 1995; Ema et al., 2010; Ito et al., 2011; Fischer et al., 2012). Rabbits are routinely used in

Table 7.9 Infectious Diseases of Rats

Pathogen	Common Names of Diseases	Fatal/Self-Limiting	Effects on Research
Parvovirus	Kilham rat virus; rat parvovirus	Asymptomatic	Immunomodulatory
Staph infection	*Staphylococcus aureus*	Dermatitis; septicemia	Significant clinical illness can result in necessity for euthanasia of the animal
Mycoplasmosis	*Mycoplasma pulmonis Mycoplasma arthritidis Mycoplasma neurolyticum Mycoplasma collis*	Rhinitis Otitis media Chronic pneumonia Infertility Can be fatal	Cause clinical disease Compromise experiments involving respiratory tract Alter immunological responses Contamination of cell lines and tumors
Cilia-associated respiratory (CAR) bacillus	*Bacillus sp.*	Respiratory disease Clinical disease—rare	
Helicobacteriosis	*Helicobacter hepaticus Helicobacter bilis, Helicobacter muridarum, Helicobacter rodentium, Helicobacter typhlonius*	IC—asymptomatic IDS— inflammatory bowel disease, diarrhea, rectal prolapse	Affects liver enzymes in some strains of rats Morbidity and mortality in rats Alter immune function
Mycobacteriosis	*Mycobacteria avium-intracellulare Mycobacteria leprae-murium*	Granulomatous pneumonia and microgranulomas in other major organs	Morbidity
Pneumocystosis	*Pneumocystis carinii*	IC—asymptomatic IDS—severe disease; pneumonia May be fatal	Pneumonia in IDS rats; may be fatal
Nematodes (Mouse pinworms)	*Syphacia obvelata Aspicularis tetraptera*	Asymptomatic with gastrointestinal lesions such as enteritis and fecal impaction	Unthriftiness in mice Alter the immune responses in animal
Acariasis (Mite Infestation)	*Myobia musculi Myocoptes musculinus Radfordia affinis*	Skin lesions, excessive pruritus (scratching), hair loss, dermatitis	Disrupt immunological responses Severe clinical disease

IC, *immune competent*; IDS, *immune compromised*.

FIGURE 7.9

New Zealand White rabbits are commonly used in research.

Photo provided by Kay Stewart.

atherosclerosis, osteoporosis, ocular, and immunology research (Southard et al., 2000; Arslan et al., 2003; McMahon et al., 2005; Castaneda et al., 2008; Habjanec et al., 2008; Manabe et al., 2008; Xiangdong et al., 2011; Panda et al., 2014; Sriram et al., 2014; Wei et al., 2014; Zhou et al., 2014). The production of polyclonal antibodies is preferentially performed in the rabbit due to its relatively large blood volume compared to rodents (Hanly et al., 1995). Their tractable nature and larger body size also make them suitable for surgical implantation of biomedical devices (Gotfredsen et al., 1995; Swindle et al., 2005; Ronisz et al., 2013). Additionally, rabbits are a favored model in pharmacologic studies for teratogenicity testing of novel pharmaceutic compounds (Gibson et al., 1966; Lloyd et al., 1999; Foote and Carney, 2000; Jiangbo et al., 2009; Oi et al., 2011).

7.3.3 NORMATIVE BIOLOGY

While much of the anatomy of the rabbit is similar to other mammalian species, it should be noted that there are a number of key differences. For example, the skin of the rabbit is quite thin and fragile. Care should be taken when restraining a rabbit or shaving a rabbit's fur (e.g., in preparation for surgery) to avoid tearing the skin. Unlike rodents and other laboratory animals, rabbits do not have pads on their feet; rather, the plantar surface is covered with fur (Quesenberry and Carpenter, 2012).

The long ears of the rabbit serve several purposes. The most obvious is for hearing. In addition, the central ear artery and marginal ear veins are easily accessible for both intravenous administration and blood sampling (Diehl et al., 2001; Parasuraman et al., 2010) (Fig. 7.10). The ears also serve as a means of thermoregulation, as excess heat may be exchanged across the large surface area of the ears (Sohn and Couto, 2012). The skin of rabbits lacks sweat glands and is therefore unable to sweat; panting is insufficient to dissipate the excess heat (Sohn and Couto, 2012). Thus, the ears play a vital role in maintaining proper body temperature. Other unique features of the skin and adnexa are the presence of chin and inguinal glands used in scent marking. Additionally, the female rabbit (doe) is noted by the presence of a large skin fold filled with subcutaneous fat just under the chin (the dewlap) (Sohn and Couto, 2012).

The skeleton of rabbits makes up only 8% of the body weight by mass (Brewer, 2006). This is in contrast to other similarly sized mammals. For example, the cat skeleton makes up 12–13% of body weight (Brewer, 2006). The small skeletal mass of the rabbit coupled with strong back muscles mean that the back is prone to traumatic fracture (Meredith and Richardson, 2015). Proper holding and restraint techniques are necessary to avoid this undesirable outcome.

There are several unique features of both the respiratory and cardiovascular systems of rabbits. For example, rabbits are obligate nose breathers (Varga, 2014). This is especially important during procedures involving anesthesia and placement of an endotracheal tube. With respect to the cardiovascular system, the rabbit heart is unique in that the right atrioventricular (AV) valve has only two leaflets instead of three (Brewer, 2006). Additionally, due to the similarity to humans with respect to the neural anatomy of the ventricles, the rabbit is the species of choice for Purkinje fiber research (Brewer, 2006).

FIGURE 7.10

Blood collection sites include the central ear artery (*black line*) and marginal ear vein (*burgundy line*) of a New Zealand White rabbit.

Photo provided by Deb Hickman.

Rabbit teeth are "open-rooted" meaning that they continue to erupt and grow throughout life. This applies to all of the teeth in the rabbit dental arcade (i.e., incisors, premolars, and molars; rabbits do not have canines). This contrasts with rodents, where the incisors are the only open-rooted (or hypsodontic) teeth (Sohn and Couto, 2012). Thus, rabbit teeth are subject to overgrowth. Another unique feature of rabbit dentition that sets them apart from rodents is the presence of a second set of incisors just behind the first set of upper incisors known as "peg" teeth (Sohn and Couto, 2012). They are thought to aid in tearing off the succulent leaves of plants while grazing.

As an obligate herbivore, the gastrointestinal tract of rabbits differs greatly from that of carnivores and omnivores. Rabbits require a high fiber diet of between 14 and 20% (Sohn and Couto, 2012). The small intestine is divided up into three main sections: the duodenum, jejunum, and ileum. The ileum connects to the cecum via a structure called the sacculus rotundus. The presence of lymph follicles suggests that the sacculus rotundus has immunological functions. It is sometimes referred to as the ileocecal "tonsil" (Jenkins, 2000). In the rabbit, gastric associated lymphoid tissue is also present in the small intestine and the vermiform appendix (Lanning et al., 2000).

The cecum, a large distensible outpouching of the large intestine, holds up to an estimated 40% of the total ingesta (Sohn and Couto, 2012). Rabbits are considered to be "hindgut fermenters." Bacteria present in the cecum ferment the digestible fiber found within the diet. The product of this fermentative process becomes cecotrophs (also known as "night feces"). Cecotrophs are excreted roughly 8 h after the initial foodstuffs are ingested (Sohn and Couto, 2012). They are softer and more mucoid in appearance than the hard, dry "day feces" produced just 4 h after consuming food (Sohn and Couto, 2012). The bulk of day feces consists of the indigestible fiber found in the diet. The sorting of foodstuffs destined to become either day feces or cecotrophs and the timing of their relative excretion is largely dependent upon the neural input of the fusus coli, also termed the "pacemaker" of the colon (Sohn and Couto, 2012). The fusus coli is anatomically located between the ascending and transverse colons of the rabbit (Popesko, 1992). Consumption of cecotrophs by rabbits is an important part of the digestive process in rabbits as they are rich in B vitamins, such as niacin and B12, and vitamin K (Hörnicke, 1981). While cecotrophs are known colloquially as "night feces," rabbits produce and eat them at all hours of the day (Sohn and Couto, 2012). Rabbits are agile enough to eat these night feces directly from their anus (Sohn and Couto, 2012). Those researchers performing digestive research (e.g., fecal collection via metabolism cages) should keep this in mind.

7.3.4 REPRODUCTIVE BIOLOGY

Sexing of adult rabbits is aided by the sexual dimorphism present in the species. Mature females are readily identified by the presence of the dewlap (Sohn and Couto, 2012). Females have 8–10 nipples, while in males these nipples are present, but rudimentary (Sohn and Couto, 2012). NZW rabbits reach sexual maturity between 5 and 7 months (Suckow et al., 2002). Reproductive data are summarized in Table 7.10.

Mature bucks have paired testicles enclosed in paired hairless scrotal sacs (Sohn and Couto, 2012). Like rodents, the inguinal rings do not close. Accessory sex glands include several bilobed organs: the seminal vesicle, vesicular gland, prostate, and paraprostatic gland. The bulbourethral glands of bucks are paired (Sohn and Couto, 2012).

Female rabbits are induced ovulators (Dal Bosco et al., 2011; Sohn and Couto, 2012). That is, the egg does not ovulate spontaneously from the ovary, rather manual stimulation via copulation is required. Ovulation occurs approximately 10 h postcopulation (Sohn and Couto, 2012). Because they are induced ovulators, does do not have a defined estrous cycle. Rather, they have periods of sexual receptivity lasting approximately 14–16 days followed by 1–2 days of nonreceptivity. Nonfertile matings may result in a period of pseudopregnancy of up to 15–16 days (Sohn and Couto, 2012). Fertile matings result in pregnancy lasting 31–32 days (Sohn and Couto, 2012). The placenta of rabbits is classified as hemodichorial; this is in contrast to humans which have a hemomonochorial placenta (Furukawa et al., 2014).

Birthing (i.e., parturition; also known as "kindling") occurs most often during the early morning hours (Sohn and Couto, 2012). Kits are born deaf and blind. By 7 and 10 days of age they can hear and see, respectively (Quesenberry and Carpenter, 2012). Amazingly, does suckle their young ones daily, usually during the dark phase, and for approximately only 5–6 min (Sohn and Couto, 2012). The kits are able to drink about 30% of their entire body weight in that time. Wild and domesticated does both follow this nursing behavior. Rabbit kits may be weaned between 5 and 8 weeks of age (Suckow et al., 2002). Earlier weaning should not be attempted as there may be profound detrimental effects on the functioning of the gastrointestinal system (Bivolarski and Vachkova, 2014).

7.3.5 BEHAVIOR

Rabbits are very social, nocturnal creatures. Scent marking is a normal and important part of their behavior repertoire. Rabbits will rub the secretions from their chin scent glands against inanimate objects, other rabbits, and human handlers in a process called "chinning" (Sohn and Couto, 2012). Dominance hierarchies are established behaviorally. Dominant animals may mount, "barber," or scent-mark subordinates (Sohn and Couto, 2012). Barbering is the act of chewing the hair of a subordinate animal, usually on the neck and back in the case of rabbits, very close to the skin giving the appearance of having been cut or "barbered" (Bays, 2006). Rabbits will "thump" one or both back feet on the ground when frightened or as an alarm call to other rabbits (Bays, 2006). Highly stressed rabbits may actually emit a loud, piercing scream, especially when roughly caught by an untrained individual (Bays, 2006). It is important to approach rabbits calmly and quietly. Relaxed,

Table 7.10 Basic Rabbit Reproductive Reference Values

Puberty	
Male	22—52 weeks
Female	22—53 weeks
Gestation	29—35 days
Litter size	4—12 kits
Weaning	5—8 weeks
Breeding duration	
Male	60—72 months
Female	24—36 months

content rabbits may be heard making a purring sound (Bays, 2006). Rabbits benefit by repeated, positive interactions with people similar to the concept of "gentling" in rats (see Section 7.2).

Changes in behavior are often first indication that an animal is in pain. Given that rabbits are a prey species, it is evolutionarily speaking not in a rabbit's best interest to display signs of pain. Thus, these behaviors are most often quite subtle in nature and may be easily missed if particular attention is not paid. The first sign often seen in a rabbit experiencing pain is a decreased appetite resulting in little to no food intake (Sohn and Couto, 2012). Rabbits will often grind their teeth (i.e., bruxism) when experiencing pain (Sohn and Couto, 2012). Other rabbits may simply appear very dull and inactive.

As with rodents, rabbits can develop stereotypies. Due to the sensitivity of the rabbit's nose and lips many stereotypies involve chewing behaviors. Bar chewing, chewing on the water bottle, and self-barbering are all stereotypical behaviors seen in rabbits (Gunn and Morton, 1995; Chu et al., 2004). In addition, rabbits may engage in "nose sliding" against solid surfaces like the cage walls and head swaying (Sohn and Couto, 2012). Animals that exhibit stereotypies do not make good research animals. Efforts should be made, where possible, to prevent these behaviors through the use of environmental enrichment. Enrichment may be in the form of chew-resistant objects (such as plastic dumbbells and stainless steel rattles) and food treats (Poggiagliolmi et al., 2011). As a prey species, rabbits benefit from the inclusion of a hut in the cage or at least a visual barrier into which they may retreat when psychologically stressed (Baumans, 2005). Breeding females should always have access to a nest box to allow for the necessary expression of normal nesting behavior (Baumans, 2005).

7.3.6 HOUSING AND MANAGEMENT

Being social creatures, ideally rabbits should be housed in compatible pairs or trios unless contraindicated by the research objectives or by incompatibility of the animals (Sohn and Couto, 2012). Stable social groups formed shortly after weaning, where animals are not added or removed, is most beneficial (Boers et al., 2002). Structurally, rabbits benefit from housing that has both adequate vertical and horizontal space (Boers et al., 2002). One recommendation on the space requirements of laboratory rabbits stipulates 16 in. as the minimum vertical cage height (National Research Council, 2011). At a minimum, rabbits must be able to comfortably sit upright in the cage without their ears bending over (National Research Council, 2011).

Laboratory rabbits are typically housed in easily sanitized stainless steel cage racks (Fig. 7.11). Slatted flooring allows for urine and feces to fall through the slats onto special pans fixed below the cage, thus providing for easier sanitation of the cages. However, care must be taken that the slats are of sufficient width so as to prevent a condition known as bumblefoot (see Section 7.3.7). Dog runs with elevated, slatted flooring or a solid floor with bedding have also been used by some investigators in group-housed rabbits with success (personal observations). Again, attention should be paid to the flooring and its effect on foot health.

Commercially available research diets specifically formulated for rabbits are available. These diets are preferred to so-called "natural diets" and feeding individual vegetables. This is because rabbits tend to be very selective eaters which can lead to nutritional imbalances (Fraser and Girling, 2009). Additionally, the use of fresh vegetables may lead to the introduction of unwanted pathogens like *Salmonella* (Varga, 2014). Commercial diets are available in maintenance and reproductive-performance dietary formulations as well as presterilized diets for rabbits housed under SPF conditions.

Rabbits are very easily heat stressed and thus must be kept at significantly lower temperatures than other laboratory animals like rats and mice. Noise is another significant stressor to rabbits (Verga et al., 2007). Sudden, high-pitched, sharp noises are most disruptive. However, in general, noise within the animal rooms should be avoided as much as possible. For this reason, rabbits should not be housed, even temporarily for short procedures, near areas of high noise.

7.3.7 DISEASES

Problems in rabbits related to the gastrointestinal system are relatively common. These problems can become serious very quickly. Therefore, it is critical that abnormalities seen (e.g., rabbit not eating or abnormal feces) be reported to the veterinary care staff immediately. Even if a researcher is unsure if there is a problem, it is best to report suspicions because without prompt intervention, seemingly minor problems can escalate to potentially life-threatening conditions.

FIGURE 7.11

Example of rabbit caging for a laboratory setting.

Photo provided by Deb Hickman.

Some of the most commonly seen clinical conditions in rabbits are summarized in Table 7.11.

7.4 ZEBRAFISH

7.4.1 INTRODUCTION

The zebrafish, *Danio rerio* of the Cyprinidae family, is a small, dark blue and yellow striped, shoaling, teleost fish, popular among aquarium enthusiasts, and increasingly among the research community (Fig. 7.12). The adult fish are 4—5 cm in length, with an incomplete lateral line and two pairs of barbels (Laale, 1977). Males have larger anal fins and more yellow coloration; females have a small genital papilla just rostral to the anal fin (Laale, 1977; Creaser, 1934).

Zebrafish are hardy, fresh water fish originating from a tropical region with an annual monsoon season. The fish are generally found among slow moving waters of rivers, streams, and wetlands, across the South Asia region of India, Bangladesh,

Table 7.11 Commonly Seen Clinical Conditions in Rabbits

Dysbiosis	Gastrointestinal upset generally secondary to the use of broad spectrum antibiotics, such as penicillin.
Malocclusion	Misalignment and subsequent overgrowth of the continuously growing teeth
Pododermatitis ("bumblefoot")	Infection of the underside of the feet
Pasteurella multocida	Common cause of respiratory infections and abscesses

FIGURE 7.12

Male zebrafish have a more stream-lined body with darker blue strips while the females have a white protruding belly.

Photo provided by Kay Stewart.

and Nepal (Engeszer et al., 2007; Spence et al., 2008). The waters tend to be shallow, relatively clear with substrates of clay, silt, or stone of varying size (McClure et al., 2006; Engeszer et al., 2007). The fish feed mostly on insects and plankton, with evidence of feeding along the water column as well as water surface (McClure et al., 2006; Engeszer et al., 2007; Spence et al., 2008).

7.4.2 USES IN RESEARCH

The small size of zebrafish, the ease of keeping large numbers, frequent spawning, large egg clutches, translucent nonadherent eggs, rapid development and complex sequencing of the zebrafish genome are all key components that make the zebrafish an attractive research model. Interestingly, approximately 70% of zebrafish genes have at least one orthologous human gene (Howe et al., 2013). Publications on the use of zebrafish in research are cited as early as the 1930s (Creaser, 1934). Until the early 1970s, the use of zebrafish stayed fairly low, with the number of articles published staying below 20 per year. In the mid-1970s, publications increased to about 40 per year, doubling again in the 1980s, increasing to almost 200 articles per year in the early 1990s, and rapidly expanding to 1929 publications by 2012.

Developmental biology was the initial focus of zebrafish research use. However, in recent years, use of the zebrafish in research related to biochemistry and molecular biology, cell biology, neurological sciences, and genetics has been rapidly increasing.

7.4.3 BIOLOGY

Zebrafish are known to live for only a year in the wild (Spence et al., 2008). For most of the year, the fish reside in shallow streams. With the onset of monsoon rains, they move to flooded, highly vegetated shallow wetlands and floodplains, including rice paddies, with little to no current and often silt bottoms for spawning (Engeszer et al., 2007). The offspring then develop in these waters until the seasonal waters diminish (Engeszer et al., 2007). Zebrafish rapidly mature, reaching sexual maturity as early as 2 months postfertilization (Lawrence et al., 2012). The zebrafish continues to grow throughout life, which is much longer in captivity, with a mean lifespan of 3.5 years in captivity (Gerhard et al., 2002).

In nature, spawning behavior occurs within small groups of three to seven fish. Males within the group pursue females, with spawning occurring along the substrate (Spence et al., 2008). Similar behaviors are noted in laboratory zebrafish, with spawning often occurring with the first light of day. Courtship behavior involves a rapid chase of the female, the male swimming around the female, nudging her, or swimming back and forth working the female to the spawning site. Interestingly, zebrafish prefer spawning near artificial plants. Once there, the male remains close to the female, extending his fins to bring his genital pore in line with the female. The male may also rapidly undulate his tail against the side of the female to initiate egg release by the female, coinciding with sperm discharge by the male. The female produces eggs in batches of 5—20 over several encounters with the male for up to an hour. Most eggs are released within the first 30 min, with a peak in production during the first 10 min (Darrow and Harris, 2004; Spence et al., 2008).

Zebrafish produce large clutches of eggs, from 150 to 400 eggs per clutch (Laale, 1977). The eggs, approximately 0.7 mn in diameter, are transparent and protected within a chorionic membrane (Kimmel et al., 1995). First body movements and beginning stages of organ development occur 10—24 h postfertilization (Kimmel et al., 1995). As development continues, the larva hatches from the egg two to three days postfertilization (Kimmel et al., 1995). The early larva has special secretory cells within multicellular regions of the head epidermis that allow the larvae to attach to various hard surfaces and plants until the swim bladder inflates 4 or 5 days postfertilization (Laale, 1977; Kimmel et al., 1995). Once the air bladder inflates, the fish can maneuver through the water column.

In captivity, zebrafish can breed year round. The presence of males, or even just the male pheromones, is needed to induce ovulation (Gerlach, 2006). If females are housed away from males for an extended period, they can retain the eggs resulting in egg-associated inflammation, which can be lethal (Kent et al., 2012a).

7.4.4 HOUSING AND MANAGEMENT

To accommodate the fish life cycle, zebrafish are typically housed in static spawning cages to allow for fertilized egg production. Spawning cages include a housing tank containing a clear slotted bottom insert, and a plastic plant. The insert is often placed in the holding tank at an angle to create a shallow region for spawning and the slotted bottom of the insert allows for ease of egg collection (Lawrence and Mason, 2012; Nasiadka and Clark, 2012) (Fig. 7.13). The embryos are then incubated at around 28.5°C in a petri dish for at least 3—4 days postfertilization (Wilson, 2012). The fish are then kept in static or slow water flow containment and can be fed *Paramecium*, rotifer, a powdered food, or a combination of these feed types.

Unfortunately, other than the need for essential fatty acids in their diet, little is yet known on the nutritional requirements of zebrafish. Zebrafish in research settings are typically fed live feed like *Artemia* (brine shrimp), rotifers, bloodworms

FIGURE 7.13

Example of a zebrafish spawning system. The system is designed to allow eggs to fall beneath a slotted insert to the bottom of the tank as a way to prevent the adult fish from consuming the eggs.

Photo provided by Robin Crisler.

(Chironomid larvae), commercial feed, or a combination of all (Lawrence, 2007). The size of the feed is necessary to suit the gape size of the larvae, approximately 100 μm (Lawrence, 2007; Wilson, 2012). Water flow and feed size increase with development, with transition of the feed to *Artemia* (brine shrimp), and/or use of a larger particle commercial feed during the 8−15 days postfertilization (Wilson, 2012).

Once the juvenile stage is reached, around 29 days postfertilization, the fish are housed more like adult fish, with more frequent feeding and slower water flow to accommodate their remaining development and smaller size, respectively (Wilson, 2012). As early as 2 months of age the fish are sexually mature (Lawrence et al., 2012).

Adult zebrafish can be housed in traditional glass aquaria or elaborate computerized and automated systems that monitor and control water quality parameters such as temperature (typically 28.5°C), pH, water hardness, salinity, dissolved oxygen, and nitrogenous wastes (Lawrence, 2007; Lawrence and Mason, 2012). Whether maintained manually or computerized, these parameters are important to monitor and maintain at appropriate levels to maximize the health of the fish. Poor water quality can lead to disease in the fish (Kent et al., 2012b).

7.4.5 DISEASES

Many of the organisms that cause disease in zebrafish are opportunists in the environment and remain subclinical until the fish is stressed, often due to problems with husbandry.

Appropriately maintained housing, combined with a healthy water quality, avoidance of overcrowding, and a functional quarantine and health surveillance program are key components to avoiding stress and disease. To date, there are currently no known viruses documented in zebrafish as naturally occurring disease concerns (Kent et al., 2012b). *Mycobacterium* infections are the most frequently documented bacterial infections (Kent et al., 2012a,b).

7.5 AMPHIBIANS AND REPTILES
7.5.1 INTRODUCTION

Class Reptilia is made up of four orders classified as Chelonia, Rhynchocephalia, Squamata, and Crocodilia (Frye, 1991). In class Amphibia, animals more commonly encountered in research setting are in the order Anura, containing frogs and toads (such as *Xenopus*, *Bufo*, *Rana*, *Hyla*, and *Dendrobates* spp.); and in the order Caudata, containing salamanders such as the tiger salamander, *Ambystoma tigrinum* and the axolotl, *Ambystoma mexicanum* (National Research Council, 1974) (Figs. 7.14 and 7.15). Snakes and lizards are in class Reptilia, order Squamata; chelonians (turtles, tortoises, terrapins) are in order Chelonia; and alligators, caimans, and crocodiles are in order Crocodilia.

FIGURE 7.14

A commonly used amphibians in research is the axolotl, *Ambystoma mexicanum.*

Provided by Chris Konz.

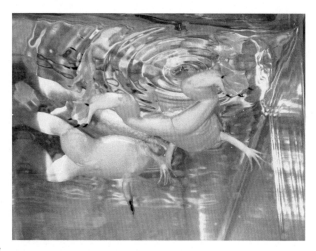

FIGURE 7.15

African clawed frogs, *Xenopus laevis*, a commonly used amphibian.

Provided by Randalyn Shepherd.

7.5.2 USES IN RESEARCH

In contrast to research in mammals, there is a tendency for reptile and amphibian research to be more oriented to studying evolution and ecology as opposed to basic science evaluating models of human disease (Pough, 1991). Salamanders and frogs are important for studying embryonic development, metamorphosis, regeneration, physiology, and climate change (Burggren and Warburton, 2007; Hopkins, 2007; Pough, 2007). Reptiles are often studied because of their more simple cardiovascular systems as well as for evaluating mechanisms of immune responses, hormonal controls, and unique reproduction methods such as parthenogenesis (Frye, 1991). Of the amphibians, *Xenopus laevis* (South African clawed frog) and *Xenopus tropicalis* (Western clawed frog) are commonly studied in the research setting. *X. laevis* is a prominent research model in comparative medicine and developmental studies, and is the most commonly studied species in the genus *Xenopus* (DeNardo, 1995; Schultz and Dawson, 2003; O'Rourke, 2007). Advantages include large-sized eggs for ease of observing embryo development, as well as the wealth of published literature in areas of research such as evolution, neurobiology, regeneration, endocrinology, and toxicology (Koustubhan et al., 2008; Gibbs et al., 2011). *Rana catesbeiana* (bullfrogs) have been used for developmental and toxicological studies, and for infectious disease study of the chytrid fungus *Batrachochytrium dendrobatidis* (Alworth and Vazquez, 2009). *A. mexicanum*, in particular, is studied to understand the regenerative ability of the blastema of amputated limbs at the molecular level (Gresens, 2004; Rao et al., 2014). *Ambystoma tigrarium* has been studied in regard to general amphibian decline in North America, environmental contaminants such as pesticides and effects of infection with *A. tigrarium* virus (Sheafor et al., 2008; Kerby and Storfer, 2009; Chen and Robert, 2011; Kerby et al., 2011). A variety of snakes, crocodiles, lizards, and turtles have been studied in research. For example, *Anolis carolinensis* (the green anole) has been used for the study of reproduction biology (Lovern et al., 2004). *Caiman crocodilus* and *Alligator mississippiensis* (crocodiles), *Trachemys scripta elegans* (red-eared sliders) represent a few other examples of reptiles used in research (O'Rourke and Schumacher, 2002).

7.5.3 BIOLOGY

Amphibians and reptiles are considered to be ectotherms (Greene, 1995). Unlike mammals and birds, ectotherms are unable to internally regulate body temperatures above that of the ambient environment through metabolism and require complex behavioral and thermoregulatory adaptations to regulate temperature (Pough, 1991; Seebacher and Franklin, 2005). In captivity, ectotherms typically require supplemental sources of heat to mimic the thermoregulatory effects of basking in the sun.

Some amphibians and reptiles are aquatic (*Xenopus* frog spp.) whereas others are semiaquatic or terrestrial. X. laevis and X. tropicalis are from geographically distinct areas and have different temperature requirement depending on life stage, with X. tropicalis adults generally around 25°C in their natural habitat versus about 20°C for X. laevis (Tinsley et al., 2010). The skin of amphibians is permeable to water and some adults (semiterrestrial tree frogs in family Hylidae, arboreal and terrestrial toads in Bufonidae) may receive a significant portion of their daily water requirement via absorption through a vascular-rich area on the pelvic area termed the pelvic patch (Pough, 2007; Ogushi et al., 2010). The skin of some amphibians contains toxins which can cause arrhythmias in human handlers, for example, alkaloids from dendrobatid frogs, and bufotoxins from toads of the genus *Bufo* (DeNardo, 1995). The toxins serve to keep predators away but, as with *Xenopus*, may harm the animals themselves by continued direct contact or diffusion through the water (Tinsley et al., 2010; Chum et al., 2013). The skin of amphibians is easily damaged thus to protect the animal during handling powder-free gloves should be worn (Gentz, 2007).

7.5.4 BEHAVIOR

Researchers and animal care providers should investigate the natural environment of each species within their care and critically evaluate what features are required for normal behavior and physiology to provide the essential elements in the research setting (Pough, 1991). In the wild, amphibians and reptiles live in ecological environments that span a range of diversity from topical forest areas to dry desert. They may be arboreal, aquatic, or terrestrial. They are often secretive and hide when in natural habitats, preferring to hide under vegetation or in crevices. Parameters from the natural habitat to evaluate include temperature, humidity, nutritional requirements, natural diet, nocturnal versus diurnal behavior, and housing density. Temperature and lighting gradients should be established so animals can choose to move toward or away from the heat source as a way to avoid overheating. Most amphibian species in the wild are nocturnal (Pough, 2007; Tinsley et al., 2010).

7.5.5 HOUSING AND MANAGEMENT

Amphibians and reptiles are sensitive to chemicals in the environment. Water quality parameters (such as pH, hardness, ammonia, nitrate/nitrite, salinity, conductivity) should be regularly monitored. Chloramine and chloramines are often present in municipal water supplies and are toxic to aquatic species. Water should be treated prior to use for aquatic species with an agent like sodium thiosulfate, since chloramine does not readily dissipate (Browne et al., 2007). Ammonia is a breakdown product between the chloramine and sodium thiosfulfate reaction and is a concern for aquatic animals (Browne et al., 2007; Koustubhan et al., 2008; O'Rourke and Schultz, 2002).

A wide variety of caging materials may be used for housing such as glass, plastic, stainless steel, or fiberglass but should be free of contaminants or harmful chemicals like bisphenol A that could leach from the caging into the water (Levy et al., 2004; Browne et al., 2007; Bhandari et al., 2015). Agents used to sanitize caging should be chosen to minimize likelihood of harmful residues.

Environmental enrichment should be provided to encourage natural behaviors and can include providing cage mates for social interaction, cage accessories that serve as hiding spots or shelters (Fig. 7.16) as well as providing a variety of food treats in changing locations for foraging opportunities (Hurme et al., 2003). Scents,

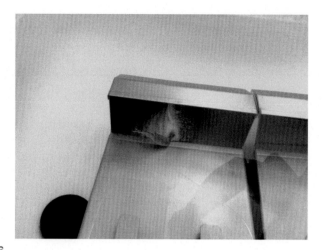

FIGURE 7.16

Use of a rabbit feeder for *Xenopus* enrichment.

Photo by Randalyn Shepherd.

sounds, and color choices may also be incorporated into enrichment strategies provided that they are carefully evaluated to ensure that they are beneficial and do not cause stress. For example, the tortoise, *Chelonoidis denticulata*, may show a color preference for red-colored enrichment items (Passos et al., 2014). PVC tubes are another example of enrichment that has been provided to *X. laevis* for use as hiding cover (Koustubhan et al., 2008). Some species may require haul-out ramps, areas for sun basking, floating rest areas, or enrichment devices along the water's surface to help prevent drowning. One should consider the possibility of ingestion, as reptiles and amphibians may attempt to consume the substrates provided to them.

The degree to which amphibians are social varies significantly depending on the species and is not always well understood. They use visual and olfactory discrimination to help them find food, forage, and avoid predators (Vitt and Caldwell, 2014). Both in the wild and in captivity, reptiles and amphibians may exhibit excitatory behavior when fed (sometimes described as a "feeding frenzy") which may result in animal injury where animals are in close proximity (Divers and Mader, 2006; Tinsley et al., 2010). Overcrowded tanks can result in competition for food and subsequent trauma. Thus, when placed together for the first time, animals should always be observed for compatibility; and only members of the same species should be housed together.

Many reptiles and amphibians are escape artists and prevention of escape and injury is a critical factor when considering housing design. Species that are prone to jumping must have secured lids on their enclosures.

The diets of amphibians and reptiles are highly variable in the wild and are species dependent. Commercially prepared pelleted diets may be available and accepted by reptiles and aquatic amphibians, however, terrestrial amphibians and many reptiles may prefer live diets (Pough, 2007). It is not unusual for some species to go several days of fasting between meals in nature (Pough, 1991). Consultation with those experienced at successful housing and feeding the species in question (zoos, nutritionists, herpetologists) is recommended.

7.5.6 DISEASES

There are many different types of infectious agents such as bacteria, viruses, fungi, and parasites that can cause health problems in amphibians and reptiles in addition to noninfectious conditions such as those resulting from nutritional imbalances, metabolic disease, neoplasia, trauma, and other spontaneous maladies. Although significant advances in knowledge have been made over the past 100 years regarding disease in these species, much still remains unknown. It is not possible to go into detail here, but there are excellent reference texts for diseases in amphibians and reptiles that can be consulted (Jacobson, 2007; Frye, 1991; Wright and Whitaker, 2001).

7.6 BIRDS

7.6.1 INTRODUCTION

From a taxonomic standpoint, birds are placed into class Aves which includes multiple orders based on anatomical, physiological, and genetic characteristics. Passeriformes is the largest order and contains songbirds and perching birds such as the finch, canary, and cardinal (Fig. 7.17). Order Columbiformes contains pigeons and doves; order Psittaciformes contains budgies and parrots such as the African gray; and order Galliformes contains domestic fowl such as the chicken and quail (Proctor and Lynch, 1993; Ritchie et al., 1994) (Fig. 7.18).

7.6.2 USES IN RESEARCH

Birds have been used as research models of human disease and are important in evaluation of aging, memory, parasitology, atherosclerosis, reproduction, and infectious disease among other topics (Austad, 1997, 2011; Maekawa et al., 2014). The genomes of several avian species have now been sequenced (Jarvis et al., 2014). Historically, chickens (*Gallus domesticus*) are the most common bird species studied in biomedical and agricultural research and are a classic model in areas such as immunology, virology, infectious disease, embryology, and toxicology (Scanes and McNabb, 2003; Kaiser, 2012). Chickens are also studied to evaluate reproductive development and retinal disease. Embryonated chicken eggs have been used to commercially produce vaccines (such as for human influenza), studied for developmental analysis, and are now being treated with viral vectors like lentivirus to produce transgenic embryos. Inbred lines with improved disease resistance are being developed and transgenic technology in the future may allow embryos to be used as bioreactors to produce therapeutic proteins of interest and potentially to generate

FIGURE 7.17

The zebra finch is a common avian species used in research.

From http://www.redorbit.com/news/science/1112751282/male-zebra-finches-fake-song-121912/.

FIGURE 7.18

The domesticated chicken commonly used in research.

Provided by Kay Stewart.

transgenic chickens which have improved resistance to pathogens (Bacon et al., 2000; Scott et al., 2010). Because chickens develop spontaneous ovarian cancers at an incidence of up to 35%, they are also a prominent model of ovarian cancer in humans (Bahr and Wolf, 2012; Hawkridge, 2014). Quail

(*Coturnix coturnix and Coturnix japonica*) have been studied in many of the same research disciplines as chickens, but offer advantages because of their smaller size and because they are among the shortest-lived bird species (Austad, 1997). Japanese quail (*C. japonica*) have been selected as a model to evaluate reproductive biology and social behaviors such as mate selection because they readily show sexual behavior in captivity (Ball and Balthazart, 2010). As with the chicken, methods to study transgenic quail are now becoming available and offer a useful tool to study gene function (Seidl et al., 2013).

Of the Psittaciformes, Amazon parrots and budgies (*Melopsittacus undulates*) are among the most commonly studied, with research topics including veterinary medicine, diagnostics, behavioral, cognition, aging, and sensory studies (Austad, 2011; Kalmar et al., 2010). The African gray parrot has been studied for its cognition and communication abilities (Hesse and Potter, 2004; Harrington, 2014).

Of the passerines studied in laboratory research, the most commonly evaluated include the zebra finch (*Taeniopygia guttata*), European starling (*Sturnus vulgaris and Sturnus roseus*), and house sparrow (*Passer domesticus*) (Bateson and Feenders, 2010). Zebra finches and other songbirds are commonly studied in regard to aging and neurogenesis in addition to speech, learning, and memory because of their ability to learn and communicate intricate bird songs (Harding, 2004; Scott et al., 2010; Austad, 2011; Mello, 2014). The most popular songbird species for neurobiological research include the zebra finch, canary, and other types of small finches such as *Lonchura striata domestica* (Schmidt, 2010). Zebra finches are favored in research settings since they are easy to house due to their small size, for their compatibility in groups, and proclivity for breeding. They are also studied for their biologic features such as sexual dimorphism, year-round singing in captivity, age-dependent period of song-learning propensity, and for ease of measurement with respect to their bird song (Fee and Scharff, 2010; Mello, 2014). Pigeons (*Columba livia*) have been evaluated in areas such as comparative psychology, neuroanatomy, neuroendocrinology, and atherosclerosis (Santerre et al., 1972; Austad, 1997; Shanahan et al., 2013). They are studied to understand their navigational skills and memory which allow homing, vision and discrimination ability. Barn owls (*Tyto alba*) are an example of a nocturnal avian species and are studied for neuroanatomy, vision, hearing, and for understanding learning mechanisms during auditory space mapping (Pena and DeBello, 2010; Rosania, 2014).

7.6.3 BIOLOGY

Birds are warm-blooded vertebrates that have feathers for the purpose of flight and plumage. Their respiratory system includes avascular air sacs, some of which attach to the lung and bronchi, but do not serve as sites for gas exchange as does the lung (Maina, 2006; Ritchie et al., 1994). Air sacs serve as internal compartments which hold air and facilitate internal air passage to allow birds to have a continuous flow of large volumes of air through the lungs as a way to increase oxygen exchange capacity and efficiency. Birds lack a functional diaphragm and use muscles of the thorax to assist with respiration (Ritchie et al., 1994). Care must be taken to ensure that use of physical restraint does not interfere with respiratory movement, cause the bird to struggle, or become stressed. The skeletal system includes pneumatic bones which are lined with air sac epithelium and are considered pneumatized by connection to the respiratory system (Frandson et al., 2009). The specific bones which are pneumatized depend on the species but typically include the humerus, cervical vertebrae, sternum, sternal ribs, and sometimes the femur (Ritchie et al., 1994).

The esophagus in birds leads to the crop, which is an outpocketing where food is held temporarily, and then continues to the proventriculus (also called the true stomach) which produces enzymes to break down food. Food travels from the proventriculus to the ventriculus (gizzard) and then on into the small and large intestines. The presence or absence of a gallbladder is species dependent (Tully et al., 2009; Kalmar et al., 2010). The rectum and urinary tract terminate in the cloaca, resulting in excreta where the fecal portion of waste is mixed with urate (white and/or creamy component). There are many additional unique and complex anatomic and physiologic adaptations of birds. Other excellent references are available in the literature (Scanes, 2015).

7.6.4 BEHAVIOR AND HOUSING

Housing requirements of birds held in captivity vary significantly depending on the particular species. Basic parameters that apply to all birds include the necessity to provide an enclosure which is safe and permits species-specific behaviors to the greatest extent possible. Consideration should be given to ensure that the type of structure is nontoxic, as some birds such as parrots have a powerful beak with the ability to chew through substrates. Enclosures may be made of metals or durable plastic, but it is important to note that zinc wire, as well as leaded paint, can be toxic to birds and is best avoided. Bar spacing on caging should be appropriate to prevent escape and injury based on the size of the bird. Caging size varies and can include large aviaries where full short-distance flight is possible, to individual housing in smaller sized cages where flight may not be feasible. Use of environmental enrichment and provision of opportunity for interaction is important to include as part of the cage structure, complexity, and social dynamic. Some types of birds are considered social, polygamous, and benefit from group housing, whereas others such as those that pair-bond (such as New World quail) may prefer housing with a single mate (Ritchie et al., 1994). Some species, genders, or individuals show aggression and may not be compatible. For example, sexually active male quail may injure each other and are generally considered incompatible (Huss et al., 2008). To help reduce aggression, housing densities should be kept low and multiple points of access to resources, such as feed and perches, should be provided.

Enrichment in the form of manipulanda can take the form of toys and food items. Some types of birds may demonstrate foraging behavior in nature and may like to manipulate their feed. Parrots, for example, typically grasp their food with their feet and may peel or strip the outer portion of the foodstuff prior to ingesting. Toys should be size appropriate for the species, easily sanitized, free from sharp edges, and replaced once wear shows. Birds can become easily caught in items that hang from the cage and as toys deteriorate they can become a hazard. For example, rope toys may begin to fray and become a hazard, causing entrapment; and some types of toys contain weights which pose a choking hazard or may be made of toxic materials such as lead. Some types of birds spend considerable time perching and require perches, which vary in diameter, for comfort and to prevent pressure sores from developing on their feet.

The respiratory system of the bird is very sensitive and caution must be taken by animal care staff to avoid exposure of birds to aerosols from chemicals that may arise from disinfectants used in the laboratory animal facility. Scented cleaners, perfumes, hairspray, and emissions from Teflon-coated materials are all examples of products which can be especially harmful to birds and may cause death.

Feeding requirements vary by species and life stage, but commercial pelleted diets designed to meet the nutritional needs can generally be provided. Although many birds are seed eaters, a diet of seeds alone is unlikely to provide adequate or balanced nutrition. Many birds have a requirement for dietary calcium, especially those that are reproductively active, and should be provided with calcium supplementation in the form of soluble grit such as cuttlebone or crushed oyster shells (Sandmeier and Coutteel, 2006; Tully et al., 2009). Birds often display neophobic behavior and may require long acclimation periods before fully accepting novel foodstuffs. For this reason, dietary changes should not be made abruptly and daily intake should be closely monitored. For birds in the laboratory setting, clean, fresh water should be provided daily either by use of nonbreakable bowls or sipper tubes. Water intake will vary by species and environmental housing conditions.

7.6.5 DISEASES

Birds can mask disease and are easily stressed. It is best to first observe the bird in its normal home environment whenever possible and only perform restraint for physical exam or collection procedures when indicated. General indications of sickness may include decreased appetite, depressed behavior, loose stools, distended abdomen, ruffled feathers or unkempt appearance, skin lesions, open-mouth breathing, abnormal respiratory sounds such as wheezing or sneezing, or signs of dehydration such as reduced skin turgor and sunken eyes. A healthy bird should have well-groomed feathers, appear alert, active and inquisitive, and should show species-typical behaviors. Its eyes should be clear and bright. No evidence of discharge should be present from the eyes, nares, mouth, or urogenital area. Numerous types of infectious (example, Fig. 7.19) and noninfectious disease presentations are described in birds. Additional reference resources should be consulted for in-depth information (Ritchie et al., 1994; Tully et al., 2009; Doneley, 2010).

FIGURE 7.19

Example of skin pox on the feet of a dark-eyed Junco (*Junco hyemalis*).

Photo from Randalyn Shepherd.

7.7 OTHER SMALL MAMMALS

To provide the reader a broader view of animal use in research, descriptions of some less commonly used small mammal models follow.

7.7.1 GUINEA PIGS

Guinea pigs (*Cavia porcellus*) are rodents, related to porcupines and chinchillas in the suborder Hystricomorpha (Fig. 7.20). They originate from the mountain and grassland regions along the mid-range of the Andes Mountains in South America. They are small, stocky, nonburrowing, crepuscular herbivores with short legs and little to no tail, ranging from 700 to 1200 g, females being smaller than males (Harkness et al., 2010). Guinea pigs have a long-standing historical role in research stretching as far back as the 1600s, when they were first used in anatomical studies (Pritt, 2012). Further, they were used by Louis Pasteur and Robert Koch in their

FIGURE 7.20

Juvenile guinea pig.

Photo from Randalyn Shepherd.

investigations of infectious disease, and have contributed to the work of several Nobel Prize worthy studies (Pritt, 2012). Specifically, the guinea pig has been used as a model for infectious diseases such as tuberculosis, Legionnaires disease, sexually transmitted diseases such as chlamydia and syphilis, and one of the more common causes of nosocomial infections in people, *Staphylococcus aureus* (Padilla-Carlin et al., 2008). Guinea pigs have also been useful tools in researching cholesterol metabolism, asthma, fetus and placental development and aspects of childbirth, as well as Alzheimer's disease (Bahr and Wolf, 2012).

Guinea pigs have many similarities to humans hormonally, immunologically, and physiologically. Unlike other rodents, and more like primates (including people), guinea pigs are prone to scurvy if they do not receive adequate vitamin C, typically in their diet (Gresham et al., 2012). Guinea pigs are housed similarly to other rodents, although they require more room than the smaller rodents.

7.7.2 HAMSTERS

Hamsters are of the *Rodentia* order, suborder *Myomorpha* along with the mouse and the rat. There are over 24 species of hamsters described in the literature, with the most common hamster used in research being the Golden or Syrian hamster, *Mesocricetus auratus* (Harkness et al., 2010) (Fig. 7.21). Originating from the northwest region of Syria, Golden hamsters are thought to be descendants of only three or four littermates collected from Syria in 1930 (Adler, 1948; Smith, 2012). As their name implies, the typical wild-type coat is reddish gold along their dorsum, with a gray underside. They are granivores and insectivores, weighing 85–150 g, females

FIGURE 7.21

Syrian hamster.

Photo from Randalyn Shepherd.

weighing more than males, with short legs and short tail, and large cheek pouches (Harkness et al., 2010).

Specific anatomical and physiological features including their susceptibility to disease and infection make them a useful model for study. Initially hamsters were utilized in studies of infectious disease, parasitology and dental disease, transitioning into cancer research in the 1960s (Smith, 2012). Hamsters are still used in many areas of research, including investigations into metabolic diseases like diabetes mellitus (Hein et al., 2013), cardiovascular disease (Russell and Proctor, 2006), reproductive endocrinology (Ancel et al., 2012), and oncology (Tysome et al., 2012). Guinea pigs have also been used as models for infectious disease associated with bacteria, parasites, and viruses, such as leptospirosis (Harris et al., 2011), leishmaniasis (Gomes-Silva et al., 2013), and severe acute respiratory syndrome (SARS) and Ebola viruses (Roberts et al., 2010; Wahl-Jensen et al., 2012).

Other species of hamsters used have been used in research. For example, Chinese and African hamsters have been used for investigations into diabetes mellitus (Kumar et al., 2012); European and Turkish hamsters have been useful to evaluate aspects of hibernation (Batavia et al., 2013); and Siberian and Turkish hamsters have been used to study circadian rhythm and pineal gland activity (Butler et al., 2008) (Fig. 7.22).

7.7.3 CHINCHILLAS

Chinchillas (Fig. 7.23) are in the order Rodentia, suborder Hystricomorpha, as are the guinea pig and the degu. There are the long-tailed chinchilla, *Chinchilla*

FIGURE 7.22

Siberian hamsters.

Photo from Greg Demas.

FIGURE 7.23

Chinchilla.

Photo from Bill Shofner Jr.

lanigera, and the short-tailed chinchilla, *Chinchilla chinchilla.* Chinchillas originate from the Andes Mountains of South America (Martin et al., 2012). They are 400−800 g in size, females weighing more than males, with compact bodies and long, strong hind limbs and dense fur coats (Alworth et al., 2012). The lushness of the coat is what led them close to extinction in the wild due to excessive hunting in the early to mid-1900s (Jimenez, 1996). The chinchilla has a large head, large eyes and ears. The large inner ear anatomy is of specific note as chinchillas are the traditional model for auditory studies (Shofner and Chaney, 2013) and otitis media (Morton et al., 2012).

7.7.4 GERBILS

The gerbil is a rodent, suborder Myomorpha, used in research. There are over 100 species of gerbil-like rodents documented, but the Mongolian gerbil (*Meriones unguiculatus*) is the species most commonly used in the United States (Fig. 7.24). Mongolian gerbils originate from a desert terrain in Mongolia and northeast China. They are long-tailed, burrowing, herbivorous rodents, 55–130 g in size, males being larger than females (Harkness et al., 2010). Due to anatomical variations in the blood supply to the brain in an anatomical region known as the "Circle of Willis," gerbils have been used most notably as a model for cerebral ischemia or stroke (Small and Buchan, 2000).

7.7.5 ARMADILLO

An interesting animal model to note among the small mammals is the nine-banded armadillo (*Dasypus novemcinctus*), a new world mammal ranging from the southeastern half of North America, extending south through the Americas to the northern region of Argentina (Balamayooran et al., 2015). Armadillos have a banded carapace, and, importantly, a low core body temperature of 33–35°C. The breeding season is in the summer, but embryo implantation is delayed until late fall, at which

FIGURE 7.24

Gerbil.

Photo used with permission of American Association for Laboratory Animal Science.

point identical quadruplicates are always formed (Balamayooran et al., 2015). The armadillo's low body temperature, and susceptibility and physiologic response to the infectious organism, *Mycobacterium leprae*, have made it an ideal model for studying leprosy (Balamayooran et al., 2015). The consistent polyembryony of the species has also made the animal a model of interest in understanding various aspects of twinning (Blickstein and Keith, 2007).

7.8 SUMMARY

Choosing the correct animal model is an essential component to the success of biomedical research. Each species used in biomedical research must be provided with adequate housing and care to ensure the well-being of the animals. Because good science and good animal care go hand in hand, it is important to understand and address the biological and behavioral needs of the animals being studied.

REFERENCES

Adler, S., 1948. Origin of the golden hamster *Cricetus auratus* as a laboratory animal. Nature 162, 256—257.

Alworth, L.C., Harvey, S.B., 2012. Chinchillas: anatomy, physiology and behavior. In: Suckow, M.A., Stevens, K.A., Wilson, R.P. (Eds.), The Laboratory Rabbit, Guinea Pig, Hamster, and Other Rodents, first ed. Academic Press/Elsevier, London.

Alworth, L.C., Vazquez, V.M., 2009. A novel system for individually housing bullfrogs. Lab. Anim. (NY) 38, 329—333.

American Fancy Rat and Mouse Association, 2014. Available from: http://www.afrma.org/.

Ancel, C., Bentsen, A.H., Sebert, M.E., Tena-Sempere, M., Mikklesen, J.D., Simonneaux, V., 2012. Stimulatory effect of RFRP-3 on the gonadotrophic axis in the male Syrian hamster: the exception proves the rule. Endocrinology 153, 1352—1363.

Arslan, H., Ketani, A., Gezici, A., Kapukaya, A., Necmioglu, S., Kesemenli, C., Subasi, M., 2003. The effects of osteoporosis on distraction osteogenesis: an experimental study in an ovariectomised rabbit model. Acta Orthop. Belg. 69, 67—73.

Austad, S.N., 1997. Birds as models of aging in biomedical research. ILAR J. 38, 137—141.

Austad, S.N., 2011. Candidate bird species for use in aging research. ILAR J. 52, 89—96.

Bacon, L.D., Hunt, H.D., Cheng, H.H., 2000. A review of the development of chicken lines to resolve genes determining resistance to diseases. Poult. Sci. 79, 1082—1093.

Baenninger, L.P., 1967. Comparison of behavioural development in socially isolated and grouped rats. Anim. Behav. 15, 312—323.

Bahr, A., Wolf, E., 2012. Domestic animal models for biomedical research. Reprod. Domest. Anim. 47 (Suppl. 4), 59—71.

Balamayooran, G., Pena, M., Sharma, R., Truman, R.W., 2015. The armadillo as an animal model and reservoir host for *Mycobacterium leprae*. Clin. Dermatol. 33, 108—115.

Ball, G.F., Balthazart, J., 2010. Japanese quail as a model system for studying the neuroendocrine control of reproductive and social behaviors. ILAR J. 51, 310—325.

Batavia, M., Nguyen, G., Zucker, I., 2013. The effects of day length, hibernation, and ambient temperature on incisor dentin in the Turkish hamster (*Mesocricetus brandti*). J. Comp. Physiol. B 183, 557–566.

Bateson, M., Feenders, G., 2010. The use of passerine bird species in laboratory research: implications of basic biology for husbandry and welfare. ILAR J. 51, 394–408.

Baumans, V., 2005. Environmental enrichment for laboratory rodents and rabbits: requirements of rodents, rabbits, and research. ILAR J. 46, 162–170.

Bays, T.B., 2006. Rabbit behavior. In: Mayer, T.B.B.L. (Ed.), Exotic Pet Behavior. W.B. Saunders, St. Louis (Chapter 1).

Benedictow, O.J., 2004. The Black Death, 1346–1353: The Complete History. Boydell Press, Rochester.

Bhandari, R.K., Deem, S.L., Holliday, D.K., Jandegian, C.M., Kassotis, C.D., Nagel, S.C., Tillitt, D.E., Vom Saal, F.S., Rosenfeld, C.S., 2015. Effects of the environmental estrogenic contaminants bisphenol A and 17alpha-ethinyl estradiol on sexual development and adult behaviors in aquatic wildlife species. Gen. Comp. Endocrinol. 214, 195–219.

Bivolarski, B.L., Vachkova, E.G., 2014. Morphological and functional events associated to weaning in rabbits. J. Anim. Physiol. Anim. Nutr. (Berl) 98, 9–18.

Blickstein, I., Keith, L.G., 2007. On the possible cause of monozygotic twinning: lessons from the 9-banded armadillo and from assisted reproduction. Twin Res. Hum. Genet. 10, 394–399.

Boers, K., Gray, G., Love, J., Mahmutovic, Z., McCormick, S., Turcotte, N., Zhang, Y., 2002. Comfortable quarters for laboratory rabbits. In: Reinhardt, V., Reinhardt, A. (Eds.), Comfortable Quarters for Laboratory Animals, ninth ed. Animal Welfare Institute, Washington, DC.

Brewer, N.R., 2006. Biology of the rabbit. J. Am. Assoc. Lab. Anim. Sci. 45, 8–24.

Browne, R.K., Odum, R.A., Herman, T., Zippel, K., 2007. Facility design and associated services for the study of amphibians. ILAR J. 48, 188–202.

Burgdorf, J., Panksepp, J., 2001. Tickling induces reward in adolescent rats. Physiol. Behav. 72, 167–173.

Burggren, W.W., Warburton, S., 2007. Amphibians as animal models for laboratory research in physiology. ILAR J. 48, 260–269.

Burman, O.H.P., Ilyat, A., Jones, G., Mendl, M., 2007. Ultrasonic vocalizations as indicators of welfare for laboratory rats (*Rattus norvegicus*). Appl. Anim. Behav. Sci. 104, 116–129.

Burt, J., 2006. Rat Reaktion Books (London).

Butler, M.P., Turner, K.W., Zucker, I., 2008. A melatonin-independent seasonal timer induces neuroendocrine refractoriness to short day lengths. J. Biol. Rhythms 23, 242–251.

Cabib, S., 2005. The neurobiology of stereotypy II: the role of stress. In: Mason, G., Rushen, J. (Eds.), Stereotypic Animal Behaviour: Fundamentals and Applications to Welfare, second ed. CABI Publishing, Cambridge, MA.

Callard, M.D., Bursten, S.N., Price, E.O., 2000. Repetitive backflipping behaviour in captive roof rats (*Rattus rattus*) and the effects of cage enrichment. Anim. Welfare 9, 139–152.

Castaneda, S., Calvo, E., Largo, R., Gonzalez-Gonzalez, R., De La Piedra, C., Diaz-Curiel, M., Herrero-Beaumont, G., 2008. Characterization of a new experimental model of osteoporosis in rabbits. J. Bone Miner. Metab. 26, 53–59.

Champneys, F., 1874. The *Septum atriorum* of the frog and the rabbit. J. Anat. Physiol. 8, 340–352.

Chen, G., Robert, J., 2011. Antiviral immunity in amphibians. Viruses 3, 2065–2086.

Chu, L.-R., Garner, J.P., Mench, J.A., 2004. A behavioral comparison of New Zealand White rabbits (*Oryctolagus cuniculus*) housed individually or in pairs in conventional laboratory cages. Appl. Anim. Behav. Sci. 85, 121–139.

Chum, H., Felt, S., Garner, J., Green, S., 2013. Biology, behavior, and environmental enrichment for the captive African clawed frog (*Xenopus spp.*). Appl. Anim. Behav. Sci. 143, 150–156.

Cloutier, S., Wahl, K., Baker, C., Newberry, R.C., 2014. The social buffering effect of playful handling on responses to repeated intraperitoneal injections in laboratory rats. J. Am. Assoc. Lab. Anim. Sci. 53, 168–173.

Creaser, C.W., 1934. The technic of handling the zebra fish (*Brachydanio rerio*) for the production of eggs which are favorable for embryological research and are available at any specified time throughout the year. Copeia 4, 159–161.

Dal Bosco, A., Rebollar, P.G., Boiti, C., Zerani, M., Castellini, C., 2011. Ovulation induction in rabbit does: current knowledge and perspectives. Anim. Reprod. Sci. 129, 106–117.

Darrow, K.O., Harris, W.A., 2004. Characterization and development of courtship in zebrafish, *Danio rerio*. Zebrafish 1, 40–45.

Denardo, D., 1995. Amphibians as laboratory animals. ILAR J. 37, 173–181.

Dickenson, V., 2013. Rabbit. Reaktion Books (London).

Diehl, K.H., Hull, R., Morton, D., Pfister, R., Rabemampianina, Y., Smith, D., Vidal, J.M., Van De Vorstenbosch, C., 2001. A good practice guide to the administration of substances and removal of blood, including routes and volumes. J. Appl. Toxicol. 21, 15–23.

Divers, S.J., Mader, D.R., 2006. Reptile Medicine and Surgery, second ed. Saunders Elsevier, St. Louis.

Doneley, B., 2010. Avian Medicine and Surgery in Practice. Manson Publishing, Ltd., London.

Einon, D.F., Morgan, M.J., 1977. A critical period for social isolation in the rat. Dev. Psychobiol. 10, 123–132.

Ema, M., Naya, M., Yoshida, K., Nagaosa, R., 2010. Reproductive and developmental toxicity of hydrofluorocarbons used as refrigerants. Reprod. Toxicol. 29, 125–131.

Engeszer, R.E., Patterson, L.B., Rao, A.A., Parichy, D.M., 2007. Zebrafish in the wild: a review of natural history and new notes from the field. Zebrafish 4, 21–38.

European Commission, 2013. Seventh Report on the Statistics on the Number of Animals Used for Experimental and Other Scientific Purposes in the Member States of the European Union. European Union, Brussels.

Fee, M.S., Scharff, C., 2010. The songbird as a model for the generation and learning of complex sequential behaviors. ILAR J. 51, 362–377.

Firth, C., Bhat, M., Firth, M.A., Williams, S.H., Frye, M.J., Simmonds, P., Conte, J.M., Ng, J., Garcia, J., Bhuva, N.P., Lee, B., Che, X., Quan, P.L., Lipkin, W.I., 2014. Detection of zoonotic pathogens and characterization of novel viruses carried by commensal *Rattus norvegicus* in New York City. MBio 5 e01933–14.

Fischer, B., Chavatte-Palmer, P., Viebahn, C., Naverrete Santos, A., Duranthon, V., 2012. Rabbit as a reproductive model for human health. Reproduction 144, 1–10.

Foote, R.H., Carney, E.W., 2000. The rabbit as a model for reproductive and developmental toxicity studies. Reprod. Toxicol. 14, 477–493.

Frandson, R.D., Wilke, W.L., Fails, A.D., 2009. Anatomy and Physiology of Farm Animals. Wiley-Blackwell, Ames.

Fraser, M.A., Girling, S., 2009. Rabbit Medicine and Surgery for Veterinary Nurses. Wiley-Blackwell, Ames.

Frye, F.L., 1991. Biomedical and Surgical Aspects of Captive Reptile Husbandry. Krieger Publishing Co., Malabar.

Furukawa, S., Kuroda, Y., Sugiyama, A., 2014. A comparison of the histological structure of the placenta in experimental animals. J. Toxicol. Pathol. 27, 11–18.

Garner, J.P., Dufour, B., Gregg, L.E., Weisker, S.M., Mench, J.A., 2004. Social and husbandry factors affecting the prevalence and severity of barbering ('whisker trimming') by laboratory mice. Appl. Anim. Behav. Sci. 89, 263–282.

Garner, J.P., Mason, G.J., 2002. Evidence for a relationship between cage stereotypies and behavioural disinhibition in laboratory rodents. Behav. Brain Res. 136, 83–92.

Gaskill, B.N., 2014. Personal Communication.

Gaskill, B.N., Gordon, C.J., Pajor, E.A., Lucas, J.R., Davis, J.K., Garner, J.P., 2012. Heat or insulation: behavioral titration of mouse preference for warmth or access to a nest. PLoS One 7, e32799.

Gaskill, B.N., Gordon, C.J., Pajor, E.A., Lucas, J.R., Davis, J.K., Garner, J.P., 2013a. Impact of nesting material on mouse body temperature and physiology. Physiol. Behav. 110–111, 87–95.

Gaskill, B.N., Karas, A.Z., Garner, J.P., Pritchett-Corning, K.R., 2013b. Nest building as an indicator of health and welfare in laboratory mice. J. Vis. Exp. 51012.

Gaskill, B.N., Pritchett-Corning, K.R., Gordon, C.J., Pajor, E.A., Lucas, J.R., Davis, J.K., Garner, J.P., 2013c. Energy reallocation to breeding performance through improved nest building in laboratory mice. PLoS One 8, e74153.

Gentz, E.J., 2007. Medicine and surgery of amphibians. ILAR J. 48, 255–259.

Gerhard, G.S., Kauffman, E.J., Wang, X., Stewart, R., Moore, J.L., Kasales, C.J., Demidenko, E., Cheng, K.C., 2002. Life spans and senescent phenotypes in two strains of Zebrafish (*Danio rerio*). Exp. Gerontol. 37, 1055–1068.

Gerlach, G., 2006. Pheromonal regulation of reproductive success in female zebrafish: female suppression and male enhancement. Anim. Behav. 72, 1119–1124.

Gibbs, K.M., Chittur, S.V., Szaro, B.G., 2011. Metamorphosis and the regenerative capacity of spinal cord axons in *Xenopus laevis*. Eur. J. Neurosci. 33, 9–25.

Gibson, J.P., Staples, R.E., Newberne, J.W., 1966. Use of the rabbit in teratogenicity studies. Toxicol. Appl. Pharmacol. 9, 398–407.

Gomes-Silva, A., Valverde, J.G., Ribeiro-Romao, R.P., Placido-Pereira, R.M., Da-Cruz, A.M., 2013. Golden hamster (*Mesocricetus auratus*) as an experimental model for *Leishmania (Viannia) braziliensis* infection. Parasitology 140, 771–779.

Gorn, E.J., Goldstein, W., 2004. A Brief History of American Sports. University of Illinois Press, Urbana.

Gotfredsen, K., Wennerberg, A., Johansson, C., Skovgaard, L.T., Hjorting-Hansen, E., 1995. Anchorage of TiO$_2$-blasted, HA-coated, and machined implants: an experimental study with rabbits. J. Biomed. Mater. Res. 29, 1223–1231.

Greene, H.W., 1995. Nonavian reptiles as laboratory animals. ILAR J. 37, 182–186.

Gresens, J., 2004. An introduction to the Mexican axolotl (*Ambystoma mexicanum*). Lab. Anim. (NY) 33, 41–47.

Gresham, V.C., Haines, V.L., 2012. Guinea pigs: managment, husbandry and colony health. In: Suckow, M.A., Stevens, K.A., Wilson, R.P. (Eds.), The Laboratory Rabbit, Guinea Pig, Hamster, and Other Rodents, first ed. Academic Press/Elsevier, London.

Gunn, D., Morton, D.B., 1995. Inventory of the behaviour of New Zealand White rabbits in laboratory cages. Appl. Anim. Behav. Sci. 45, 277–292.

Habjanec, L., Halassy, B., Vdovic, V., Balija, M.L., Tomasic, J., 2008. Comparison of mouse and rabbit model for the assessment of strong PGM-containing oil-based adjuvants. Vet. Immunol. Immunopathol. 121, 232−240.

Hans, J.S., Bird, G.C., Li, W., Jones, J., Neugbauer, V., 2005. Computerized analysis of audible and ultrasonic vocalizations of rats as a standardized measure of pain-related behavior. J. Neurosci. Meth. 141, 261−269.

Hanly, W.C., Artwohl, J.E., Bennett, B.T., 1995. Review of polyclonal antibody production procedures in mammals and poultry. ILAR J. 37, 93−118.

Harding, C.F., 2004. Learning from bird brains: how the study of songbird brains revolutionized neuroscience. Lab. Anim. (NY) 33, 28−33.

Hardy, C., Callou, C., Vigne, J.D., Casane, D., Dennebouy, N., Mounolou, J.C., Monnerot, M., 1995. Rabbit mitochondrial DNA diversity from prehistoric to modern times. J. Mol. Evol. 40, 227−237.

Harkness, J.E., Turner, P.V., VandeWoude, S., Wheler, C.L., 2010. Harkness and Wagner's Biology and Medicine of Rabbits and Rodents, fifth ed. Wiley, Ames.

Harrington, M., 2014. Speaking of psittacine research. Lab. Anim. (NY) 43, 343.

Harris, B.M., Blatz, P.J., Hinkle, M.K., McCall, S., Beckius, M.L., Mende, K., Robertson, J.L., Griffith, M.E., Murray, C.K., Hospenthal, D.R., 2011. In vitro and in vivo activity of first generation cephalosporins against *Leptospira*. Am. J. Trop. Med. Hyg. 85, 905−908.

Hawkridge, A.M., 2014. The chicken model of spontaneous ovarian cancer. Proteomics Clin. Appl. 8, 689−699.

Hedrich, H.J., 2000. History, strains, and models. In: Krinke, G. (Ed.), The Laboratory Rat. Academic Press, San Diego.

Hein, G.J., Baker, C., Hsieh, J., Farr, S., Adeli, K., 2013. GLP-1 and GLP-2 as yin and yang of intestinal lipoprotein production: evidence for predominance of GLP-2-stimulated postprandial lipemia in normal and insulin-resistant states. Diabetes 62, 373−381.

Hesse, B.E., Potter, B., 2004. A behavioral look at the training of Alex: a review of Pepperberg's the Alex studies: cognitive and communicative abilities of grey parrots. Anal. Verbal Behav. 20, 141−151.

Hickman, D.L., Swan, M., 2010. Use of a body condition score technique to assess health status in a rat model of polycystic kidney disease. J. Am. Assoc. Lab. Anim. Sci. 49, 155−159.

Hirsjarvi, P.A., Junnila, M.A., Valiaho, T.U., 1990. Gentled and non-handled rats in a stressful open-field situation; differences in performance. Scand. J. Psychol. 31, 259−265.

Hopkins, J., Bourdain, A., Freeman, M., 2004. Extreme Cuisine: The Weird & Wonderful Foods that People Eat. Tuttle Publishing, North Clarendon.

Hopkins, W.A., 2007. Amphibians as models for studying environmental change. ILAR J. 48, 270−277.

Hori, M., Yamada, K., Ohnishi, J., Sakamoto, S., Furuie, H., Murakami, K., Ichitani, Y., 2014. Tickling during adolescence alters fear-related and cognitive behaviors in rats after prolonged isolation. Physiol. Behav. 131, 62−67.

Hörnicke, H., 1981. Utilization of caecal digesta by caecotrophy (soft faeces ingestion) in the rabbit. Livestock Prod. Sci. 8, 361−366.

Howe, K., Clark, M.D., Torroja, C.F., Torrance, J., Berthelot, C., Muffato, M., Collins, J.E., Humphray, S., McLaren, K., Matthews, L., McLaren, S., Sealy, I., Caccamo, M., Churcher, C., Scott, C., Barrett, J.C., Koch, R., Rauch, G.J., White, S., Chow, W., Kilian, B., Quintais, L.T., Guerra-Assuncao, J.A., Zhou, Y., Gu, Y., Yen, J., Vogel, J.H.,

Eyre, T., Redmond, S., Banerjee, R., Chi, J., Fu, B., Langley, E., Maguire, S.F., Laird, G.K., Lloyd, D., Kenyon, E., Donaldson, S., Sehra, H., Almeida-King, J., Loveland, J., Trevanion, S., Jones, M., Quail, M., Willey, D., Hunt, A., Burton, J., Sims, S., McLay, K., Plumb, B., Davis, J., Clee, C., Oliver, K., Clark, R., Riddle, C., Elliot, D., Threadgold, G., Harden, G., Ware, D., Begum, S., Mortimore, B., Kerry, G., Heath, P., Phillmore, B., Tracey, A., Corby, N., Dunn, M., Johnson, C., Wood, J., Clark, S., Pelan, S., Griffiths, G., Smith, M., Glithero, R., Howden, P., Barker, N., Lloyd, C., Stevens, C., Harley, J., Holt, K., Panagiotidis, G., Lovell, J., Beasley, H., Henderson, C., Gordon, D., Auger, K., Wright, D., Collins, J., Raisen, C., Dyer, L., Leung, K., Robertson, L., Ambridge, K., Leongamornlert, D., McGuire, S., Gilderthorp, R., Griffiths, C., Manthravadi, D., Nichol, S., Barker, G., et al., 2013. The zebrafish reference genome sequence and its relationship to the human genome. Nature 496, 498−503.

Hurme, K., Gonzalez, K., Halvorsen, M., Foster, B., Moore, D., Chepko-Sade, B.D., 2003. Environmental enrichment for dendrobatid frogs. J. Appl. Anim. Welf. Sci. 6, 285−299.

Huss, D., Poynter, G., Lansford, R., 2008. Japanese quail (*Coturnix japonica*) as a laboratory animal model. Lab. Anim. (NY) 37, 513−519.

Ibi, D., Takuma, K., Koike, H., Mizoguchi, H., Tsuritani, K., Kuwahara, Y., Kamei, H., Nagai, T., Yoneda, Y., Nabeshima, T., Yamada, K., 2008. Social isolation rearing-induced impairment of the hippocampal neurogenesis is associated with deficits in spatial memory and emotion-related behaviors in juvenile mice. J. Neurochem. 105, 921−932.

Ito, T., Ando, H., Handa, H., 2011. Teratogenic effects of thalidomide: molecular mechanisms. Cell. Mol. Life Sci. 68, 1569−1579.

Jacobson, E.R., 2007. Infectious Diseases and Pathology of Reptiles: Color Atlas and Text. CRC Press, Boca Raton.

Jarvis, E.D., Mirarab, S., Aberer, A.J., Li, B., Houde, P., Li, C., Ho, S.Y., Faircloth, B.C., Nabholz, B., Howard, J.T., Suh, A., Weber, C.C., Da Fonseca, R.R., Li, J., Zhang, F., Li, H., Zhou, L., Narula, N., Liu, L., Ganapathy, G., Boussau, B., Bayzid, M.S., Zavodovych, V., Subramanian, S., Gabaldon, T., Capelle-Gutierrez, S., Huerta-Cepas, J., Rekepalli, B., Munch, K., Schierup, M., Lindow, B., Warren, W.C., Ray, D., Green, R.E., Bruford, M.W., Zhan, X., Dixon, A., Li, S., Li, N., Huang, Y., Derryberry, E.P., Bertelsen, M.F., Sheldon, F.H., Brumfield, R.T., Mello, C.V., Lovell, P.V., Wirthlin, M., Schneider, M.P., Prosdocimi, F., Samaniego, J.A., Velazquez, A.M.V., Alfaro-Nunez, A., Campos, P.F., Petersen, B., Sicheritz-Ponten, T., Pas, A., Bailey, T., Scofield, P., Bunch, M., Lambert, D.M., Zhou, Q., Perelman, P., Driskell, A.C., Sharpiro, B., Xiong, Z., Zeng, Y., Liu, S., Li, Z., Liu, B., Wu, K., Xiao, J., Yinqi, X., Zheng, Q., Zhang, Y., Yang, H., Wang, J., Smeds, L., Rheindt, F.E., Braun, M., Fjeldsa, J., Orlando, L., Barker, F.K., Jonsson, K.A., Johnson, W., Koepfli, K.P., O'Brien, S., Haussler, D., Ryder, O.A., Rahbek, C., Willerslev, E., Graves, G.R., Glenn, T.C., McCormack, J., Burt, D., Ellegren, H., Alstrom, P., Edwards, S.V., Stamatakis, A., Mindell, D.P., Cracraft, J., et al., 2014. Whole-genome analyses resolve early branches in the tree of life of modern birds. Science 346, 1320−1331.

Jenkins, J.R., 2000. Rabbit and ferret liver and gastrointestinal testing. In: Fudge, A.M. (Ed.), Laboratory Medicine: Avian and Exotic Pets. Saunders, Philadelphia.

Jiangbo, Z., Xuying, W., Yuping, Z., Xili, M., Yiwen, Z., Tianbao, Z., 2009. Effect of astragaloside IV on the embryo-fetal development of Sprague-Dawley rats and New Zealand White rabbits. J. Appl. Toxicol. 29, 381−385.

Jimenez, J.E., 1996. The extirpation and current status of wild chinchillas *Chinchilla lanigera* and *C-brevicaudata*. Biol. Conserv. 77, 1–6.

Jourdan, D., Ardid, D., Chapuy, E., Eschalier, A., Le Bars, D., 1995. Audible and ultrasonic vocalization elicited by single electrical nociceptive stimuli to the tail in the rat. Pain 63, 237–249.

Kaiser, P., 2012. The long view: a bright past, a brighter future? Forty years of chicken immunology pre- and post-genome. Avian Pathol. 41, 511–518.

Kalmar, I.D., Janssens, G.P., Moons, C.P., 2010. Guidelines and ethical considerations for housing and management of psittacine birds used in research. ILAR J. 51, 409–423.

Kent, M.L., Harper, C., Wolf, J.C., 2012a. Documented and potential research impacts of sub-clinical diseases in zebrafish. ILAR J. 53, 126–134.

Kent, M.L., Spitsbergen, J.M., Matthews, J.M., Fournie, J.W., Westerfield, M., 2012b. Diseases of Zebrafish in Research Facilities. https://zebrafish.org/wiki/health/disease_manual/start.

Kerby, J.L., Hart, A.J., Storfer, A., 2011. Combined effects of virus, pesticide, and predator cue on the larval tiger salamander (*Ambystoma tigrinum*). Ecohealth 8, 46–54.

Kerby, J.L., Storfer, A., 2009. Combined effects of atrazine and chlorpyrifos on susceptibility of the tiger salamander to *Ambystoma tigrinum* virus. Ecohealth 6, 91–98.

Kimmel, C.B., Ballard, W.W., Kimel, S.R., Ullmann, B., Schilling, T.F., 1995. Stages of embryonic development of the zebrafish. Dev. Dyn. 203, 253–310.

Koustubhan, P., Sorocco, D., Levin, M.S., Conn, P.M., 2008. Establishing and maintaining a *Xenopus laevis* colony for research laboratories. In: Conn, P.M. (Ed.), Sourcebook of Models for Biomedical Research. Humana Press, Totowa.

Kumar, S., Singh, R., Vasudeva, N., Sharma, S., 2012. Acute and chronic animal models for the evaluation of anti-diabetic agents. Cardiovasc. Diabetol. 11, 9.

Laale, H.W., 1977. The biology and use of zebrafish, *Brachydanio rerio* in fisheries research. A literature review. J. Fish Biol. 10, 121–176.

Lanning, D., Zhu, X., Zhai, S.K., Knight, K.L., 2000. Development of the antibody repertoire in rabbit: gut-associated lymphoid tissue, microbes, and selection. Immunol. Rev. 175, 214–228.

Laulederkind, S.J., Hayman, G.T., Wang, S.J., Smith, J.R., Lowry, T.F., Nigam, R., Petri, V., De Pons, J., Dwinell, M.R., Shimoyama, M., Munzenmaier, D.H., Worthey, E.A., Jacob, H.J., 2013. The rat genome database 2013–data, tools and users. Brief Bioinform. 14, 520–526.

Lawrence, C., 2007. The husbandry of zebrafish (*Danio rerio*): a review. Aquaculture 269, 1–20.

Lawrence, C., Adatto, I., Best, J., James, A., Maloney, K., 2012. Generation time of zebrafish (*Danio rerio*) and medakas (*Oryzias latipes*) housed in the same aquaculture facility. Lab. Anim. (NY) 41, 158–165.

Lawrence, C., Mason, T., 2012. Zebrafish housing systems: a review of basic operating principles and considerations for design and functionality. ILAR J. 53, 179–191.

Levy, G., Lutz, I., Kruger, A., Kloas, W., 2004. Bisphenol A induces feminization in *Xenopus laevis* tadpoles. Environ. Res. 94, 102–111.

Lindsey, J.R., Baker, H.J., 2006. Historical perspectives. In: Suckow, M.A., Weisbroth, S.H., Franklin, C.L. (Eds.), The Laboratory Rat, second ed. Elsevier, Boston.

Lloyd, M.E., Carr, M., McElhatton, P., Hall, G.M., Hughes, R.A., 1999. The effects of methotrexate on pregnancy, fertility and lactation. QJM 92, 551–563.

Lovern, M.B., Holmes, M.M., Wade, J., 2004. The green anole (*Anolis carolinensis*): a reptilian model for laboratory studies of reproductive morphology and behavior. ILAR J. 45, 54−64.

Maekawa, F., Tsukahara, S., Kawashima, T., Nohara, K., Ohki-Hamazaki, H., 2014. The mechanisms underlying sexual differentiation of behavior and physiology in mammals and birds: relative contributions of sex steroids and sex chromosomes. Front. Neurosci. 8, 242.

Maina, J.N., 2006. Development, structure, and function of a novel respiratory organ, the lung-air sac system of birds: to go where no other vertebrate has gone. Biol. Rev. Camb. Philos. Soc. 81, 545−579.

Manabe, Y.C., Kesavan, A.K., Lopez-Molina, J., Hatem, C.L., Brooks, M., Fujiwara, R., Hochstein, K., Pitt, M.L., Tufariello, J., Chan, J., McMurray, D.N., Bishai, W.R., Dannenberg Jr., A.M., Mendez, S., 2008. The aerosol rabbit model of TB latency, reactivation and immune reconstitution inflammatory syndrome. Tuberculosis (Edinb) 88, 187−196.

Martin, B.J., 2012. Chinchillas: taxonomy and history. In: Suckow, M.A., Stevens, K.A., Wilson, R.P. (Eds.), The Laboratory Rabbit, Guinea Pig, Hamster, and Other Rodents, first ed. Academic Press, London.

McClure, M.M., McIntyre, P.B., McCune, A.R., 2006. Notes on the natural diet and habitat of eight danionin fishes, including the zebrafish *Danio rerio*. J. Fish Biol. 69, 553−570.

McMahon, A.C., Kritharides, L., Lowe, H.C., 2005. Animal models of atherosclerosis progression: current concepts. Curr. Drug Targets Cardiovasc. Haematol. Disord. 5, 433−440.

Mello, C.V., 2014. The zebra finch, *Taeniopygia guttata*: an avian model for investigating the neurobiological basis of vocal learning. Cold Spring Harb. Protoc. 2014, 1237−1242.

Meredith, A.L., Richardson, J., 2015. Neurological diseases of rabbits and rodents. J. Exot. Pet Med. 24, 21−33.

Morton, D.J., Hempel, R.J., Seale, T.W., Whitby, P.W., Stull, T.L., 2012. A functional tonB gene is required for both virulence and competitive fitness in a chinchilla model of *Haemophilus influenzae* otitis media. BMC Res. Notes 5, 327.

Nasiadka, A., Clark, M.D., 2012. Zebrafish breeding in the laboratory environment. ILAR J. 53, 161−168.

National Research Council, 2011. Guide for the Care and Use of Laboratory Animals, eighth ed. National Academies Press, Washington, DC.

National Research Council (U.S.), 1974. Amphibians: Guidelines for the Breeding, Care, and Management of Laboratory Animals; a Report. National Academies Press, Washington, DC.

O'Rourke, D.P., 2007. Amphibians used in research and teaching. ILAR J. 48, 183−187.

O'Rourke, D.P., Schultz, T.W., 2002. Biology and diseases of amphibians. In: Fox, J.G., Anderson, L.C., Loew, F.M., Quimby, F.W. (Eds.), Laboratory Animal Medicine, second ed. Academic Press, Amsterdam.

O'Rourke, D.P., Schumacher, J., 2002. Biology and diseases of reptiles. In: Fox, J.G., Anderson, L.C., Loew, F.M., Quimby, F.W. (Eds.), Laboratory Animal Medicine, second ed. Academic Press, Amsterdam.

Ogushi, Y., Tsuzuki, A., Sato, M., Mochida, H., Okada, R., Suzuki, M., Hillyard, S.D., Tanaka, S., 2010. The water-absorption region of ventral skin of several semiterrestrial and aquatic anuran amphibians identified by aquaporins. Am. J. Physiol. Regul. Integr. Comp. Physiol. 299, R1150−R1162.

Oi, A., Morishita, K., Awogi, T., Ozaki, A., Umezato, M., Fujita, S., Hosoki, E., Morimoto, H., Ishiharada, N., Ishiyama, H., Uesugi, T., Miyatake, M., Senba, T., Shiragiku, T., Nakagiri, N., Ito, N., 2011. Nonclinical safety profile of tolvaptan. Cardiovasc. Drugs Ther. 25 (Suppl. 1), S91−S99.

Padilla-Carlin, D.J., McMurray, D.N., Hickey, A.J., 2008. The Guinea pig as a model of infectious diseases. Comp. Med. 58, 324−340.

Panda, A., Tatarov, I., Masek, B.J., Hardick, J., Crusan, A., Wakefield, T., Carroll, K., Yang, S., Hsieh, Y.H., Lipsky, M.M., McLeod, C.G., Levine, M.M., Rothman, R.E., Gaydos, C.A., Detolla, L.J., 2014. A rabbit model of non-typhoidal *Salmonella bacteremia*. Comp. Immunol. Microbiol. Infect. Dis. 37, 211−220.

Panksepp, J., 2007. Neuroevolutionary sources of laughter and social joy: modeling primal human laughter in laboratory rats. Behav. Brain Res. 182, 231−244.

Parasuraman, S., Raveendran, R., Kesavan, R., 2010. Blood sample collection in small laboratory animals. J. Pharmacol. Pharmacother. 1, 87−93.

Parasana, A.J., Li, N., Brown, T.H., 2012. Positive and negative ultrasonic social signals elicit opposing firing patterns in rat amygdala. Behav. Brain Res. 226, 77−86.

Passos, L.F., Mello, H.E., Young, R.J., 2014. Enriching tortoises: assessing color preference. J. Appl. Anim. Welf. Sci. 17, 274−281.

Patterson-Kane, E.G., 2003. Shelter enrichment for rats. Contemp. Top. Lab. Anim. Sci. 42, 46−48.

Pena, J.L., Debello, W.M., 2010. Auditory processing, plasticity, and learning in the barn owl. ILAR J. 51, 338−352.

Poggiagliolmi, S., Crowell-Davis, S.L., Alworth, L.C., Harvey, S.B., 2011. Environmental enrichment of New Zealand White rabbits living in laboratory cages. J. Vet. Behav. Clin. Appl. Res. 6, 343−350.

Popesko, P., Rajtová, V., Horák, J.R., 1992. A Colour Atlas of the Anatomy of Small Laboratory Animals. Wolfe Publishing, London.

Pough, F.H., 1991. Recommendations for the care of amphibians and reptiles in Academic Institutions. ILAR J. 33, S1−S21.

Pough, F.H., 2007. Amphibian biology and husbandry. ILAR J. 48, 203−213.

Pritt, S., 2012. Guinea pigs: taxonomy and history. In: Suckow, M.A., Stevens, K.A., Wilson, R.P. (Eds.), The Laboratory Rabbit, Guinea Pig, Hamster, and Other Rodents, first ed. Academic Press, London.

Proctor, N.S., Lynch, P.J., 1993. Manual of Ornithology: Avian Structure & Function. Yale University Press, New Haven.

Quesenberry, K.E., Carpenter, J.W., 2012. Ferrets, Rabbits, and Rodents: Clinical Medicine and Surgery. Elsevier/Saunders, St. Louis.

Rao, N., Song, F., Jhamb, D., Wang, M., Milner, D.J., Price, N.M., Belecky-Adams, T.L., Palakal, M.J., Cameron, J.A., Li, B., Chen, X., Stocum, D.L., 2014. Proteomic analysis of fibroblastema formation in regenerating hind limbs of *Xenopus laevis* froglets and comparison to axolotl. BMC Dev. Biol. 14, 1−27.

Rappuoli, R., 2014. Inner workings: 1885, the first rabies vaccination in humans. Proc. Natl. Acad. Sci. U.S.A. 111, 12273.

Ritchie, B.W., Harrison, G.J., Harrison, L.R., 1994. Avian Medicine: Principles and Application. Wingers Publishing, Lake Worth.

Robbins, T.W., Jones, G.H., Wilkinson, L.S., 1996. Behavioural and neurochemical effects of early social deprivation in the rat. J. Psychopharmacol. 10, 39−47.

Roberts, A., Lamirande, E.W., Vogel, L., Baras, B., Goossens, G., Knott, I., Chen, J., Ward, J.M., Vassilev, V., Subbarao, K., 2010. Immunogenicity and protective efficacy in mice and hamsters of a beta-propiolactone inactivated whole virus SARS-CoV vaccine. Viral Immunol. 23, 509–519.

Ronisz, A., Delcroix, M., Quarck, R., 2013. Measurement of right ventricular pressure by telemetry in conscious moving rabbits. Lab. Anim. 47, 175–183.

Rosania, K., 2014. Barn owls: why give a hoot? Lab. Anim. (NY) 43, 157.

Roy, C.S., 1879. The form of the pulse-wave: as studied in the carotid of the rabbit. J. Physiol. 2, 66–90, 11.

Russell, J.C., Proctor, S.D., 2006. Small animal models of cardiovascular disease: tools for the study of the roles of metabolic syndrome, dyslipidemia, and atherosclerosis. Cardiovasc. Pathol. 15, 318–330.

Russell, L.D., 1992. Normal development of the testes. In: Mohr, U., Dungworth, D.L., Capen, C.C. (Eds.), Pathobiology of the Aging Rat. ILSI Press, Washington, DC.

Sandmeier, P., Coutteel, P., 2006. Management of canaries, finches, and mynahs. In: Harrison, G.J., Lightfoot, T.L. (Eds.), Clinical Avian Medicine, vol. 2. Spix Publishing, Palm Beach.

Santerre, R.F., Wight, T.N., Smith, S.C., Brannigan, D., 1972. Spontaneous atherosclerosis in pigeons. A model system for studying metabolic parameters associated with atherogenesis. Am. J. Pathol. 67, 1–22.

Scanes, C.G., 2015. Sturke's Avian Physiology. Elsevier, London.

Scanes, C.G., McNabb, F.M.A., 2003. Avian models for research in toxicology and endocrine disruption. Avian Poult. Biol. Rev. 14, 21–52.

Schmidt, M.F., 2010. An IACUC perspective on songbirds and their use in neurobiological research. ILAR J. 51, 424–430.

Schultz, T.W., Dawson, D.A., 2003. Housing and husbandry of *Xenopus* for oocyte production. Lab. Anim. (NY) 32, 34–39.

Scott, B.B., Velho, T.A., SIM, S., LOIS, C., 2010. Applications of avian transgenesis. ILAR J. 51, 353–361.

Seebacher, F., Franklin, C.E., 2005. Physiological mechanisms of thermoregulation in reptiles: a review. J. Comp. Physiol. B 175, 533–541.

Seidl, A.H., Sanchez, J.T., Schecterson, L., Tabor, K.M., Wang, Y., Kashima, D.T., Poynter, G., Huss, D., Fraser, S.E., Lansford, R., Rubel, E.W., 2013. Transgenic quail as a model for research in the avian nervous system: a comparative study of the auditory brainstem. J. Comp. Neurol. 521, 5–23.

Shanahan, M., Bingman, V.P., Shimizu, T., Wild, M., Gunturkun, O., 2013. Large-scale network organization in the avian forebrain: a connectivity matrix and theoretical analysis. Front. Comput. Neurosci. 7, 1–17.

Sheafor, B., Davidson, E.W., Parr, L., Rollins-Smith, H.,L., 2008. Antimicrobial peptide defenses in the salamander, *Ambystoma tigrinum*, against emerging amphibian pathogens. J. Wildl. Dis. 44, 226–236.

Shiomi, M., Ito, T., 2009. The Watanabe heritable hyperlipidemic (WHHL) rabbit, its characteristics and history of development: a tribute to the late Dr. Yoshio Watanabe. Atherosclerosis 207, 1–7.

Shiomi, M., Ito, T., Yamada, S., Kawashima, S., Fan, J., 2003. Development of an animal model for spontaneous myocardial infarction (WHHLMI rabbit). Arterioscler Thromb. Vasc. Biol. 23, 1239–1244.

Shofner, W.P., Chaney, M., 2013. Processing pitch in a nonhuman mammal (*Chinchilla laniger*). J. Comp. Psychol. 127, 142−153.

Small, D.L., Buchman, A.M., 2000. Animal models. Br. Med. Bull 56, 307−317.

Smith, G.D., 2012. Hamsters: taxonomy and history. In: Suckow, M.A., Stevens, K.A., Wilson, R.P. (Eds.), The Laboratory Rabbit, Guinea Pig, Hamster, and Other Rodents. Academic Press, London.

Smith, W.R., 1891. Abnormal arrangement of the right subclavian artery in a rabbit. J. Anat. Physiol. 25, 325−326.

Sohn, J., Couto, M.A., 2012. Anatomy, physiology, and behavior. In: Suckow, M.A., Stevens, K.A., Wilson, R.P. (Eds.), The Laboratory Rabbit, Guinea Pig, Hamster, and Other Rodents. Academic Press, Boston.

Song, Y., Lan, Z., Kohn, M.H., 2014. Mitochondrial DNA phylogeography of the Norway rat. PLoS One 9.

Southard, T.E., Southard, K.A., Krizan, K.E., Hillis, S.L., Haller, J.W., Keller, J., Vannier, M.W., 2000. Mandibular bone density and fractal dimension in rabbits with induced osteoporosis. Oral Surg. Oral Med. Oral Pathol. Oral Radiol. Endod. 89, 244−249.

Spence, R., Gerlach, G., Lawrence, C., Smith, C., 2008. The behaviour and ecology of the zebrafish, *Danio rerio*. Biol. Rev. 83, 13−34.

Sriram, S., Gibson, D.J., Robinson, P., Pi, L., TuliI, S., Lewin, A.S., Schultz, G., 2014. Assessment of anti-scarring therapies in ex vivo organ cultured rabbit corneas. Exp. Eye Res. 125, 173−182.

Suckow, M.A., Brammer, D.W., Rush, H.G., Chrisp, C.E., 2002. Biology and diseases of rabbits. In: Fox, J.G., Anderson, L.C., Loew, F.M., Quimby, F.W. (Eds.), Laboratory Animal Medicine, second ed. Academic Press, Amsterdam.

Swindle, M.M., Nolan, T., Jacobson, A., Wolf, P., Dalton, M.J., Smith, A.C., 2005. Vascular access port (VAP) usage in large animal species. Contemp. Top. Lab. Anim. Sci. 44, 7−17.

Thomas, D., Mayhew, H., 1998. The Victorian Underworld. New York University Press, New York.

Tinsley, H.R., 2010. Amphibians, with special reference to *Xenopus*. In: Hubrecht, R., Kirkwood, J. (Eds.), The UFAW Handbook on the Care and Management of Laboratory and Other Research Animals, eighth ed. Wiley-Blackwell, Ames.

Tully, T.N., Dorrestein, G.M., Jones, A.K., 2009. Handbook of Avian Medicine, second ed. Elsevier, Oxford.

Tysome, J.R., Li, X., Wang, S., Wang, P., Gao, D., Du, P., Chen, D., Gangeswaran, R., Chard, L.S., Yuan, M., Alusi, G., Lemoine, N.R., Wang, Y., 2012. A novel therapeutic regimen to eradicate established solid tumors with an effective induction of tumor-specific immunity. Clin. Cancer Res. 18, 6679−6689.

Ullman-Cullere, M.H., Foltz, C.J., 1999. Body condition scoring: a rapid and accurate method for assessing health status in mice. Lab. Anim. Sci. 49, 319−323.

Van Bergeijk, J.P., Van Herck, H., De Boer, S.F., Meijer, G.W., Hesp, A.P., Van Der Gugten, J., Beynen, A.C., 1990. Effects of group size and gentling on behaviour, selected organ masses and blood constituents in female Rivm: TOX rats. Z. Versuchstierkd 33, 85−90.

Varga, M., 2014. Textbook of Rabbit Medicine. Elsevier Butterworth Heinemann, Edinburgh.

Verga, M., Luzi, F., Carenzi, C., 2007. Effects of husbandry and management systems on physiology and behaviour of farmed and laboratory rabbits. Horm. Behav. 52, 122−129.

Vitt, L.J., Caldwell, J.P., 2014. Herpetology: An Introductory Biology of Amphibians and Reptiles. Academic Press, Amsterdam.

Wahl-Jensen, V., Bollinger, L., Safronetz, D., De Kok-Mercado, F., Scott, D.P., Ebihara, H., 2012. Use of the Syrian hamster as a new model of ebola virus disease and other viral hemorrhagic fevers. Viruses 4, 3754–3784.

Wei, C., Zhu, M., Petroll, W.M., Robertson, D.M., 2014. *Pseudomonas aeruginosa* infectious keratitis in a high oxygen transmissible rigid contact lens rabbit model. Invest. Ophthalmol. Vis. Sci. 55, 5890–5899.

Whishaw, I.Q., Kolb, B., 2005. The Behavior of the Laboratory Rat: A Handbook With Tests. Oxford University Press, Oxford.

Wilson, C., 2012. Aspects of larval rearing. ILAR J. 53, 169–178.

Wooding, F.B.P., Burton, G., 2008. Comparative Placentation: Structures, Functions and Evolution. Springer, Berlin.

Wright, K.M., Whitaker, B.R., 2001. Amphibian Medicine and Captive Husbandry. Krieger Publishing, Malabar.

Xiangdong, L., Yuanwu, L., Hua, Z., Liming, R., Qiuyan, L., Ning, L., 2011. Animal models for the atherosclerosis research: a review. Protein Cell 2, 189–201.

Zhou, Q., Liu, X.Y., Ruan, Y.X., Wang, L., Jiang, M.M., Wu, J., Chen, J., 2014. Construction of corneal epithelium with human amniotic epithelial cells and repair of limbal deficiency in rabbit models. Hum. Cell 28, 22–36.

Zinsser, H., 1935. Rats, Lice, and History. George Rutledge & Sons, London.

Common Technical Procedures in Rodents

K.L. Stewart
University of Notre Dame, Notre Dame, IN, United States

CHAPTER OUTLINE

8.1 HANDLING AND RESTRAINT METHODS

Studies have revealed that even minimal handling of laboratory animals can cause stress (Balcombe et al., 2004). Fortunately, animals typically become accustomed to repeated proper handling. Manual restraint of mice and rats often begins with grasping them at the **base** of the tail and gently lifting them. If an animal, especially a large rat, is grasped by the tip of the tail, the weight of the animal can cause the skin of the tail to deglove or fall off. Suspending the animal in the air for an extended period of time must be avoided; they should be held only long enough to transfer them into another cage during cage changing or onto another surface for other procedures. If suspended too long, the animal can experience distress and even injury if it twists around and tries to flip itself upright.

The most common method of restraint for mice and young rats is scruffing them by the fur at the back of the neck (AALAS Learning Library, 2005a). Older rats are carefully grasped around the shoulders with minimal pressure placed on the chest of the animal (Fig. 8.1). Rats are rib breathers; they must be able to expand their rib cage to breathe. If restrained too tightly, they cannot breathe (AALAS Learning Library, 2005b; Machholz et al., 2012).

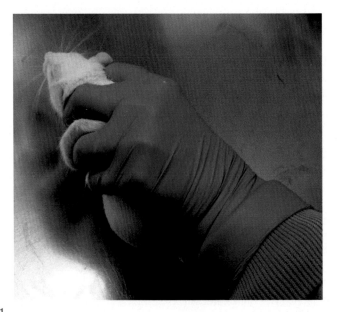

FIGURE 8.1

The proper manual restraint of an adult rat utilizing the "T-Rex" method. The hand is placed over the shoulder with the index finger on one side of the head and the middle finger on the other side. The fingers on each side of the head restrict side-to-side movement of the head. Encircle the body behind the forelegs with the third and fourth fingers and the thumb.

Numerous types of restraint devices are commercially available for use on animals of all sizes. Plastic restraint tubes are available in a variety of sizes for mice, rats, and other rodents. Quality restraint devices prevent the animal from turning around and allow easy access to strategic parts of the animal. Devices should also be easy to sanitize and provide adequate ventilation.

8.2 IDENTIFICATION

Cage cards are utilized to capture information such as animal identity, the strain of animal, sex, number of animals in the cage, principal investigator, and the Institutional Animal Care and Use Committee protocol number. Cage cards must always remain with the cage to avoid misidentification of the animals. Temporary identification of individual animals can be accomplished by nontoxic pen markings on the tail, hair clipping, or dyeing the fur. Pen marks will only last a few days whereas hair clipping may last up to 14 days. Ear punch identification and ear tags can be utilized on smaller animals but may be obliterated by fighting between individuals. Microchips and tattoos have also been used for identification (Robinson et al., 2003).

8.3 COMMON TECHNICAL PROCEDURES

As the animal protocols often include the administration of a test compound or the collection of blood for sampling, it is prudent that the correct techniques are used. As stated earlier, studies have demonstrated that even minimal handling of mice and rats can be stressful. The act of moving an animal from one cage to another is shown to cause an increase in heart rate, blood pressure, and other physiological parameters, such as serum corticosterone, with these fluctuations continuing up to several hours (Balcombe et al., 2004). The changes in blood parameters can alter the data collected. Thus, technical procedures must be done by trained individuals with approved methods to minimize the occurrence of stress and discomfort to the research animals.

8.3.1 BLOOD WITHDRAWAL TECHNIQUES

Many experimental protocols require blood sampling. The method of collection is dependent on the species, the volume needed, and the skill level of the technician. In mice and rats, the approximate volume of circulating blood is 55–70 mL/kg of body weight (Diehl et al., 2001). Generally, up to 10% of the circulating blood volume can be taken on a single occasion from normal healthy animals with minimal adverse effect. This volume may be repeated after 2–3 weeks. For repeated blood sampling at shorter intervals, a maximum of 1% of an animal's circulating blood volume can be removed every 24 h. However, for animals that are not physiologically normal, care should be taken in these calculations, as the percentage of circulating blood will be about 15% lower in obese and older animals. If too much blood is withdrawn too rapidly or too frequently without fluid replacement, hypovolemic shock and/or anemia can occur (Diehl et al., 2001).

Common sites of blood withdrawal in mice and rats are the retro-orbital sinus and the tail vein (via amputation of the tail tip); both require that the animal be anesthetized. The submandibular vein of the mouse and the saphenous vein in both mice and rats can be used as bleeding sites in conscious animals. To harvest a large volume of blood from rodents, an intracardiac puncture can be performed. However, this is a terminal procedure that requires the animal to be fully anesthetized.

8.3.1.1 Retro-Orbital

The use of the retro-orbital sinus or vessels is a blood collection technique that can be performed on mice, hamsters, rats, gerbils, and guinea pigs. This procedure should only be performed by a skilled technician. The animal must be anesthetized for the procedure. With the animal positioned in lateral recumbency, the head is secured between the thumb and the forefinger. A microhematocrit tube is placed into the medial canthus of the eye. Gentle pressure is applied on the tube while it is rotated on its axis as it reaches the back of orbit. Once the tube has cut through the membrane, blood will begin to flow into and through the tube (Fig. 8.2). The

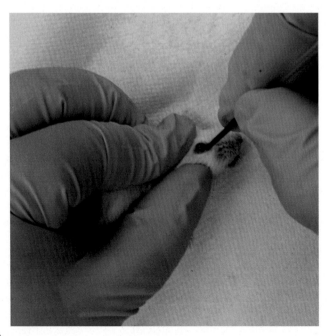

FIGURE 8.2

For the retro-orbital bleeding method, the anesthetized animal is placed in lateral recumbency and the head is held such that the skin is pulled taut causing the eye to protrude.

tube is removed when it is completely filled. To prevent the blood from spilling out of the tube, a finger is placed over one end. The tube is sealed at the other end with sealing clay. Once the tube is removed from the eye, the bleeding should cease. Assistance with hemostasis is accomplished by closing the eye lids and applying gentle pressure with a cotton ball. Retro-orbital sampling has been largely replaced with less-invasive blood collection techniques such as from the lateral saphenous vein or submandibular vein.

8.3.1.2 Facial and Submandibular Veins

In the mouse, there are two vessels on the face that can be used for blood collection: the facial vein, which runs from the ocular plexus across the cheek and the submandibular vein, which runs along the lower jaw. The vessels converge into the jugular vein. Blood collection from these vessels does not require anesthetizing the animal and can be performed several times without harm to the animal. However, neither side of the face should be used more often than every 5–7 days (Joslin, 2009). For the collection of blood from the facial vein and submandibular veins, proper restraint of the animal is essential. If the grip on the animal is too tight, the blood flow to the veins can be restricted. However, the head must have very restricted movement.

Blood collection from the submandibular vein requires the use of a needle that is inserted no more than 4–5.5 mm to prevent trauma to the muscles, nerves, and other vessels that are in the head, neck, and oral cavity. A lancet can also be used for the puncture of the vessels. The choice of the lancet size, ranging from 4 to 5.5 mm, is dependent on the size of the animal with the 5.5 mm used on animals over 20 g in size. The approximate area of the submandibular vein is located by the point where a line from the corner of the mouth intersects a line from the lateral canthus of the eye. This coincides with a small hairless dimple found caudal to the corner of the mouth and slightly below the jaw line.

While restraining the mouse, the area of the facial vein is located by measuring the length of the eye below the lateral canthus and the width of the eye caudally. The point of the lancet is used to gently feel for the location at which the jaw bone ends. The lancet is then pushed firmly into the skin and vessel.

For both procedures, once the lancet or needle is removed, the blood will flow from the puncture. Drops of blood are collected into a small tube. Once sufficient blood has been collected, the scruff of the animal should be released to stop the blood flow.

8.3.1.3 Tail Vein

The distal tail vein procedure can be used to collect blood from mice and rats. The distal end of the tail is cleaned with warm water. Alcohol should not be used as it will cause vasoconstriction. Warming the tail before the procedure with a warm wash cloth or a heating pad will increase blood flow. The distal tip of the tail is amputated with a sterile surgical blade or a sharp pair of sterile surgical scissors. However, no more than 1 mm of tail tissue should be removed from a mouse or 2 mm from a rat

(NIH.gov, 2010). The blood collection tube is positioned under the clipped part of the tail. Stroking the tail or squeezing the tail from the base to the distal end can increase blood flow but can also increase the contamination of the sample with other cells and other tissue products. Once sufficient blood has been collected, direct pressure is applied to the tip of the tail using a cotton ball for 20–30 s until hemostasis is confirmed. Repeated samples can be taken by the gentle removal of the clot at the end of the tail.

The lateral tail vein procedure is also used to collect blood from mice and rats. The animal is prepared as described for the distal vein collection. The tail is held between the forefinger and thumb to stabilize it. Gentle pressure is placed on the tail. A needle is used to pierce the skin, entering at a very shallow angle, almost parallel to the tail. The needle is advanced until a "flash" of blood is seen in the hub of the needle. The blood must be aspirated slowly to avoid collapsing the vessel. Once the sample is collected, the needle is removed and direct pressure is applied over the venipuncture site for 20–30 s until hemostasis has been established.

8.3.1.4 Intracardiac Puncture

When a study requires a larger volume of blood, the intracardiac collection method is often used. Approximately half of the total blood volume can be collected from a mouse or rat via these methods, 40 μL/g or approximately 1 mL for an average 25 g mouse and approximately 10 mL of blood for a 250 g rat. This is a terminal procedure that requires general anesthesia (Adeghe and Cohen, 1986). The intracardiac method can be performed using either the closed method or by surgically exposing the heart and inserting the needle directly into the left ventricle (open method). For the closed method, the landmarks for the needle placement are the groove formed by the rib cage at the xiphoid process, on the animal's left side. The animal is placed in dorsal recumbency or is held with the nondominant hand as vertical as possible. The needle is inserted in the space, just lateral to the midline, at a 15–20-degree angle. It is then advanced and directed slightly toward the left while aspirating the plunger at the same time. The needle is advanced until a "flash" of blood is observed in the hub of the needle. Aspirate the blood slowly until the sample has been collected. The needle is removed once the flow has ceased.

8.3.1.5 Lateral Saphenous and Cephalic Veins

Both the lateral saphenous and cephalic vein procedures are used to collect blood from rats and mice (Hem et al., 1998). Clear plastic flexible restraint cones can be used for the restraint of the animal. The hair is removed at the venipuncture site (Fig. 8.3). The vein is occluded just above the site using a tourniquet. Using a 22-gauge needle, the vessel is punctured at a 30-degree angle and the blood is collected directly into hematocrit tubes. Hemostasis through application of gentle pressure with a gauze pad should be achieved prior to releasing the animal from the restraint device.

FIGURE 8.3

Saphenous vein bleeding procedure. The hair at the puncture is shaved to prevent the blood from pooling in the follicles and to minimize contamination of the sample.

8.3.2 COMPOUND ADMINISTRATION TECHNIQUES

Preparation for administration of compounds to research animals includes determining the proper dosage and the proper route of administration. Because many of the compounds used are produced in the laboratory, there must be assurance that the compound or solution is sterile if it is to be injected and physiologically compatible. For proper absorption and to prevent tissue damage, all compounds must be physiologically buffered to the proper pH. The viscosity must also be considered, as a solution that is too thick to pass through a small gauge needle may require reformulation for oral administration (Turner et al., 2011a,b).

The use of the appropriate equipment, such as the smallest gauge needle, and appropriate restraint devices will minimize tissue trauma and discomfort to the animal. The choice of the needle size is based on the route of administration, the viscosity of the solution, and the size of the animal. The smallest size feasible to administer the solution, usually 22–30 gauge in the mouse and 20–25 gauge for the rat should be used. Injection volumes range from 0.1 mL to a maximum of 0.3 mL per injection site depending on the route used (Turner et al., 2011a,b).

Injections are commonly given intramuscularly (IM), subcutaneously (SQ), intravenously (IV), intradermally (ID), or intraperitoneally (IP) (Sirois, 2016). Compounds can also be administered orally, through the use of a gavage needle that deposits the material directly into the stomach or via feed that has the compound incorporated into it. Knowledge of the anatomy of the species is imperative as an improperly placed injection can cause muscle, tissue, or nerve damage, especially with intramuscular injections of rodents. The absorption rate of compounds varies in accordance with the route. When a substance is administered IV, there is no absorption required as it is placed directly into the blood stream. When a substance is injected IM, it is rapidly absorbed due to the abundant number of vessels within the muscle tissue. Because of the small body size of the mouse and rat, the SQ dosing route is the only convenient and safe way to administer a large volume of fluid. This method allows for a sustained effect of the solution as the absorption rate is slower than other routes (Smith, 1993).

When preparing the material for injection, the stopper top on the drug bottle should be cleaned with alcohol before withdrawing the dose. The drug is withdrawn slowly from the bottle to minimize air in the syringe. Because air bubbles injected IV can potentially cause an air embolus, all air bubbles must be removed from the syringe by holding the syringe upright and tapping it. The air will rise to the top of the syringe and can then be expelled from the syringe. Aseptic technique must be followed with the injection of materials by using sterile, disposable syringes and needles. Before a solution is injected, the placement of the needle must be confirmed by slightly pulling back the plunger of the syringe to create a vacuum. This is known as **aspiration**. If blood is aspirated during a subcutaneous or intramuscular injection, the placement of the needle is incorrect.

8.3.2.1 Intramuscular Injection

Proper needle size is imperative for a successful intramuscular injection; for mice, a 25–27 gauge needle and for rats a 22–25 gauge needle is suggested. Typical IM injection volumes for mice range from 0.01 to 0.1 mL for the gluteal muscle and up to 0.05 mL for the gastrocnemius muscle. For the rat, injection volumes typically range from 0.01 to 0.3 mL for the gluteal muscle and up to 0.1 mL for the gastrocnemius. The animal should be securely restrained with the injection site easily accessible. This is accomplished either with manual restraint performed by an assistant or with the use of a restraint device. The hind leg of both the mouse and rat leg are small with the gluteal muscles (the thigh muscles) comprising the largest muscle mass in the hind limb. The second largest muscle mass in the hind limb is the gastrocnemius (the calf muscle). When the hind limb is extended by grasping the skin of the flank at the cranial portion of the femur, the stifle is prevented from bending. The gluteal muscle mass can then be palpated. The midline of the leg from the posterior aspect runs from the point of the hock to the tail. There is a ridge created in the hair that grows in opposite direction from the inner and outer thigh, meeting at the midline of the leg. Using this ridge as a landmark, the injection site should be toward the lateral aspect of the thigh, just off the midline, reducing the possibility of damaging the nerves and vessels that are located on the medial surface of the thigh.

To ensure minimal discomfort of the animal, extraneous movement of the needle in the tissue must be avoided. Proper restraint of the animal and proper positioning of the technician's hand prevent the need to reposition the hand when performing the injection. The syringe must be aspirated to ensure placement within the muscle and not in a blood vessel. The material is injected in a steady fluid motion. If injections are made too rapidly, the muscle tissue expands, resulting in tissue trauma and pain to the animal.

Because of the small muscle mass of many rodents, an intramuscular injection may cause discomfort and local tissue irritation, especially if too large a volume of a solution or a solution with an overly acidic or alkaline pH is administered. An understanding of anatomy and careful technique are necessary to avoid the sciatic nerve in the hind leg, just behind the femur. Injection into or close to the nerve may lead to unnecessary discomfort, temporary lameness, or permanent paralysis of the leg. As a result of nerve damage, an animal might chew on the affected extremity.

8.3.2.2 Subcutaneous Injection

Subcutaneous administration utilizes the space that is created between the skin layers and the muscle when the skin at the nape of the neck is lifted. Needle size is determined by the viscosity of the material to be injected, generally a 22–30-gauge needle for mice and 22–25-gauge needle for rats. Injection volumes range from 0.1 to 0.5 mL for mice and 0.1 to 1.0 mL for rats.

Manual restraint of the mouse is performed by grasping the animal by the scruff of the neck and placing it on a solid surface with the heel of the hand resting on the surface. Care must be taken to avoid placing excessive pressure on the animal as that can inhibit breathing. For rats, it is advisable to use a restraint device to avoid injury to the technician. The skin at the nape of the neck is grasped upward to form a tent of the skin. If using a restraint device, forceps may be required to grasp the skin through a slot that is present on the top of some restraint devices. The needle is inserted parallel to the spine and directed away from the head at the base of the skin fold of the tent (Fig. 8.4). The needle is directed away from the head to avoid the possibility of injury to the skull of the mouse as it is very thin and to avoid accidental injection into the rat neck muscles as rats have a tendency to rear their heads back.

Aspiration is required to check for proper needle placement. If the needle is properly placed, there will be back pressure on the syringe when the plunger is pulled back. The injection should be done with a slow steady motion. When completed, the needle can be slightly rotated prior to removal and the skin pinched at the injection site to prevent loss of injected material.

8.3.2.3 Intravenous Injection

Although intravenous injection is the most direct route of substance administration to introduce compounds into the circulatory system, it can be challenging

FIGURE 8.4

The proper technique for a subcutaneous injection of the mouse. Note that the needle is directed away from the head to prevent injury to the skull.

to perform in mice and rats due to the small vessels and the thickness of the tail skin. If repeated intravenous administration is required, the use of vascular access ports or other specialized dosing equipment should be considered for the welfare of the animals. Needle selection should be the smallest size possible that will allow for the viscosity of the material injected, generally 27–30 gauge. Injection volumes range from 0.05 mL to a maximum of 1.0 mL per injection based on the size of the animal, with lower volumes used in mice compared to rats.

Dilation of the tail veins by warming the tail prior to the injection procedure will facilitate the success of this procedure. Heating the tail can be achieved by submerging the tail into warm water of approximately 47°C for 1 min, wrapping warm towels around the tail for 1 min, or wrapping it or the whole body of the animal for a short time in a heating pad that is on the warm or medium setting. As the tail is warmed, the vessels will dilate and can be seen when the tip of the tail is lifted and rotated slightly in either direction. The needle is inserted almost parallel to the vein, as the vein is very superficial. If the vein has been penetrated, blood will flash in the needle hub. The tip of the needle can be followed visually as it penetrates the vein. Accurate placement of the needle into the vessel can be confirmed when the vessel is visually flushed as the compound is administered. The formation of a bleb at the site indicates that the needle is intradermal, between skin layers. If necessary, a second attempt can be performed by removing the needle and trying a site on the same vessel in a more proximal location on the tail.

8.3.2.4 Intradermal Injection

Intradermal injections are delivered into the outer layers of the dermis, underneath the upper skin layer (the epidermis). This procedure requires that the animal remain very still; anesthesia is often needed so that animals are sufficiently restrained.

Most intradermal injections are aqueous-based compounds that are physiologically buffered to have a neutral pH. If the solution is not buffered, tissue necrosis can occur at the injection site. The dose range per injection site is 50−100 μL. Injections exceeding this range can cause tissue necrosis at the injection site or leakage of the compound out of the site due to pressure. The needle size range is typically 25−30 gauge.

The injection site is prepared by clipping the hair at the injection site. The skin is stretched taut between the technician's thumb and index finger, providing stability to the skin when positioning the needle. The needle is placed, bevel up, on the skin and is gently inserted into the skin between the epidermis and the dermis layers. The needle is advanced just beyond the bevel and the substance is injected slowly. A small bleb or blister will form in the skin, indicating that the needle was properly placed. When removing the needle, allow the skin to stretch over the bleb to prevent loss of the injected material. Blotting or wiping the area should be avoided as that can cause the compound to leak from the injection sites.

8.3.2.5 Intraperitoneal Injection

The intraperitoneal route for the injection of compounds into rodents is commonly employed because it can be used for the delivery of larger volumes than possible via the intravenous or intramuscular routes. However, the absorption of material that is administered IP is significantly slower than for the IV route. Typical needle sizes are 22−30 gauge for mice and 22−25 gauge for rats. Injection volumes for mice range from 0.05 to 1.0 mL based on the size of the mouse and 0.1−1.5 mL for rats.

This procedure can be easily done solo for mice but usually requires two people for rats. The mouse is grasped by the scruff of skin at the nape of the neck and the hand is adjusted to ensure that the mouse cannot turn its head. The hind quarters can be secured by holding the tail between the third and fourth fingers. For the rat, one person restrains the rat and the other one performs the injection. The mouse or rat is positioned to expose the abdomen, facing it upward. The head is tilted downward at about a 30-degree angle to allow the intestines to fall forward.

The landmarks for the injection can be identified as follows: the cranial border is an imaginary line extending horizontally across the body at the top of the hip (from flank to flank), the medial border is the line that is created from the hair growth from opposite directions meeting, and the lateral border line is created from the top of the hip to the prepuce in the male and from the top of the hip and following the teats in the female. These borders provide a triangular area that is safe to use as the injection site. The needle is inserted perpendicular to the spine off the midline in the triangle as described previously (Fig. 8.5). At this angle, the needle will "pop" into the abdomen allowing for easier determination of depth, and it is a visual as well as tactile cue that the needle is properly positioned. Aspiration of the syringe is

FIGURE 8.5

The intraperitoneal injection is performed with the use of manual restraint with the animal positioned at a 30-degree angle, the head tilting downward. The needle is inserted perpendicular to the spine. Note the black stopper plug to the right in the figure that has been modified to serve as a needle holder; the cap of the needle is secured in the holder and allows safe placement of the needle back in the cap after use, thus reducing the risk of needle-stick injury.

required to assure that the needle is in the cavity and not in the intestines, urinary bladder, or a blood vessel. Following the injection, the needle is pulled out at the 90 degree angle to prevent trauma within the peritoneal cavity.

8.3.2.6 Gavage

Oral gavage is a precise method of enteral dosing, as it places the compound directly into the stomach of the animal at a specific time and at a specified volume. Typically, commercially available stainless steel–feeding needles are used to deliver compound by this method. The length of the feeding needle required can be estimated by determining the distance from the tip of the nose to last rib. Common feeding needle diameters range from 20 to 22 gauge for mice and 18 to 20 gauge for rats. The gavage needle is placed onto the syringe with the graduations on the syringe barrel visible without turning the needle. Once the needle has been placed into the esophagus, any rotation can rupture the esophagus resulting in death of the animal. The volume for gavage of an animal is normally 10 mL/kg. The maximum volume is 40 mL/kg, however, large gavage doses have been shown to overload the stomach capacity and pass immediately into the small bowel with a possibility of reflux into the esophagus.

Proper manual restraint of the animal is crucial to the success of this technique as injury to the oral cavity, pharynx, larynx, trachea, esophagus, and stomach can result from improper positioning of the head and the body. The body must be held so that it is suspended in a straight line from the head to the tail (Fig. 8.6). Any twisting of the body will impede the placement of the gavage needle into the esophagus. This procedure should be performed with the animal conscious as a way to decrease the possibility of placing the needle into the trachea.

The tip of the needle is placed in the rear of the animal's mouth to induce swallowing, while the shaft of the needle is gently pressed against the roof of the mouth to extend the neck and align the mouth and esophagus. The tip is then slid down back of mouth, the tip moving forward in one fluid motion. Any resistance felt indicates improper placement; the gavage needle should slide down into esophagus easily. A violent reaction (coughing, gasping) indicates accidental introduction of the tube into the larynx or trachea. Once the needle is properly placed, the compound should be slowly administered. Once the compound has been administered, the needle should be removed without rotation.

8.4 RODENT ANESTHESIA

8.4.1 INDUCTION

The goal of the anesthetist is to induce and maintain an animal at the desired anesthesia level using the minimal amount of anesthesia. The induction of anesthesia for

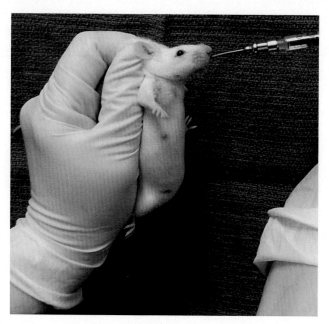

FIGURE 8.6

To gavage a mouse, the proper restraint method is crucial to the success of the procedure. Once the needle is introduced into the animal's mouth, it is used to lift the head upward to allow the gavage needle to slide into the esophagus.

rodents is accomplished either with an injectable anesthetic or an inhalant anesthetic. Injectable anesthetics can also be used for the initial induction while maintenance is accomplished through the use of an inhalant anesthesia, usually isoflurane or sevoflurane (Tsukamoto, 2015). Inhalant anesthetics can be administered by use of a precision vaporizer, thus making it easier to regulate the depth of anesthesia. Maintenance of the surgical plane of anesthesia utilizing inhalant anesthesia is done by adjusting the percent concentration of anesthetic being delivered through a precision vaporizer. The precision vaporizer can deliver a precise concentration of anesthetic gas to animals placed in an induction chamber (Fig. 8.7). During the induction of the anesthesia utilizing an isoflurane vaporizer, the vaporizer setting is 3–4%, whereas for maintenance of anesthesia the levels are reduced to 1–1.5% (Cesarovic et al., 2010). For anesthetic maintenance the animal can be fitted with a nose cone that is either directly attached to the anesthesia system or is part of a manifold that provides multiple anesthetic stations so that several animals can be anesthetized simultaneously.

Alternatively, an anesthetic-soaked cotton ball can be placed in a bell jar beneath a grid or platform on which the animal stands; and as the anesthetic vaporizes into the air within the jar, the animal becomes anesthetized. The disadvantage of using the bell jar is that it is impossible to control the anesthetic concentration, making overdose more likely. As well, once the animal is removed from the bell jar, it may begin to regain consciousness. In either case, the waste anesthetic gas must be either filtered through a charcoal filter or directly vented out of the room via a fume hood to prevent exposure of personnel.

The most commonly used injectable anesthetic for rodents is ketamine in combination with sedatives or muscle relaxers. Because of the high concentration of the

FIGURE 8.7

An induction chamber can be used to induce anesthesia in an animal using an inhalant anesthetic. The chamber is in-line with the precision vaporizer and the exhaust is directed into a charcoal filter canister.

base compounds, dilutions of the drug combinations are usually prepared as a stock from which doses for individual animals can be drawn. All doses are based on the strain of the animal, the age, and the overall health status. Common combinations include the following:

1. Ketamine (100 mg/mL), xylazine (20 mg/mL), acepromazine (10 mg/mL) diluted with sterile saline (0.9% NaCl). This combination is dosed according to body weight (BW) of the mouse using the following calculation: $(BW \times 10) - 50 =$ microliters of rodent cocktail to be given. This mixture can be administered IM or IP.
2. Ketamine–xylazine 2:1 that consists of ketamine (100 mg/mL) and xylazine (20 mg/mL) diluted with sterile saline (0.9% NaCl). This combination is primarily used for anesthesia of rats and can be used in conjunction with inhalation anesthesia. For adult rats, an effective initial dose is typically 0.3 cc; however, rats that are required to undergo subsequent anesthetic events may require increased doses of approximately 0.02 cc each time because they may develop the ability to metabolize the anesthetic agent more quickly. This anesthetic combination should be generally administered by the IM route.
3. Ketamine–xylazine "Mouse Mix" that consists of ketamine (100 mg/mL) and xylazine (20 mg/mL) diluted with sterile saline (0.9% NaCl). This combination is again given according to the weight of the mouse using the following calculation: $(BW \times 10) - 50 =$ microliters. This can be administered IP.

The main disadvantage of the use of ketamine is that it is a Drug Enforcement Administration–controlled substance that requires that a controlled substance license be obtained.

8.4.2 STAGES OF ANESTHESIA

There are four stages of anesthesia and four planes within the surgical stage of anesthesia. During stage I, the animal simply becomes disoriented while trying to investigate its surroundings. Stage II is the excitatory phase. The animal will exhibit an irregular respiration rate that can include the holding of its breath in some mice and rats strains. The ability to roll back over when placed in a dorsal position, known as the righting reflex, is lost during this phase.

It is at Stage III that the surgical stage of anesthesia is achieved. There are four planes within this stage of anesthesia and various reflexes of the animal can be used to evaluate anesthetic plane as described in Section 8.4.3. In Plane I, the palpebral and swallowing reflexes are absent, while the corneal reflex is lost in Plane II. During these two planes there are no amnesia or analgesic effects. It is at Plane III that the animal reaches the surgical level of anesthesia. Plane III includes paralysis of the intercostal muscles, which then results in diaphragmatic respiration. Initially there is only partial analgesia in Plane III, but it ultimately advances into complete amnesia and analgesia as the anesthesia level deepens. It is only at this depth of anesthesia that a surgical procedure can be initiated.

If the anesthesia level further deepens, there are complications that can result in death of the animal. Should an animal reach Plane IV of the third stage of anesthesia, it has been overdosed and can progress quickly into Stage IV. In this stage there is complete paralysis of both the intercostal muscles and the diaphragm resulting in severe apnea. Respiratory arrest, medullary paralysis, and vasomotor collapse follow and finally death occurs.

8.4.3 ASSESSMENT OF ANESTHETIC DEPTH

The depth of anesthesia is assessed by testing the response to various stimuli. If voluntary movement results from physical stimuli of the body, the animal is not at the anesthetic level required for surgery. The swallowing reflex manifests as an attempt by the animal to swallow normal salivary secretions. Physical methods utilized for anesthetic depth assessment include the toe pinch, tail pinch, ear pinch, palpebral reflex, and corneal reflex. To use the toe pinch method, the leg is extended and the webbing between the toes is isolated. Using finger nails or atraumatic forceps, the area is firmly pinched. If the leg is retracted or the foot is withdrawn, it is a positive reaction to the stimulus. The tail pinch and ear pinch utilize the same method, but on the tip of the tail or the pinnae of the ear. Any movement of the tail or ears indicates a positive reaction to the stimulation. For assessment of the palpebral reflex, a fingertip is touched to the medial canthus (inner corner) of the eye, and for the corneal reflex, a sterile cotton-tipped applicator is used to gently touch the cornea (eyeball). Blinking or movement of the whiskers is indicative of a positive response. Generally, the animal is not at a surgical plane of anesthesia if there is movement, vocalization, or marked increase in respirations.

Physiological indicators such as heart rate, respiratory rate, blood pressure, mucous membrane color, and capillary refill time can also be used to evaluate depth of anesthesia. General observations are useful to detect changes in the respiratory rate of the animals. However, specialized equipment is required to determine the heart rate or blood pressure. The color of the mucous membranes, color of the eyes, ears, mouth, nose, anus, and to a lesser extent the paws and tail can be observed. All of these areas should be pink, indicating adequate respiration and cardiac function. However, if the animal moves to Stage IV anesthesia, cyanosis of the mucous membranes and surrounding skin will develop. Capillary refill time, the amount of time taken for color to return to an external capillary bed after it has been blanched, is tested by the application of gentle pressure over an area. For example, an applicator stick or a finger can be pressed on the gums, pinnae, or nail beds of the anesthetized animals. It should take no more than $1-2$ s for the blanched area to return to its normal pink color. An extended refill time can indicate a reduction in heart rate or strength of cardiac contraction that is caused by the animal being too deeply anesthetized and, possibly, near death.

It is important to use multiple parameters to assess anesthetic depth. Areas tested for a pinch reflex will become desensitized if used repeatedly. Using both physical and physiological parameters provide a complete evaluation of the anesthetized animal. Anesthetic depth should be reassessed every 10–30 min throughout the surgical procedure.

8.5 RODENT SURGERY

The *Guide for the Care and Use of Laboratory Animals* (the Guide) (Institute for the Laboratory Animal Research, 2011) recommends that rodent survival surgery be performed aseptically. Minimizing the chances for contamination of the surgical site is accomplished through the use of specific practices including patient preparation, surgeon preparation, sterilization of instruments and other supplies, and the use of a clean and controlled environment. Preparation of the patient, proper preparation of the surgeon, and the use of sterile gloves and instruments will be discussed in this section. A primary cause of rodent surgical death is hypothermia. Steps to avoid exacerbating hypothermia are also discussed.

8.5.1 SURGICAL PATIENT PREPARATION

Prior to performing the surgical preparation, the animal must be anesthetized. It is positioned to allow access to the surgical site. For mice and rats, positioning for abdominal procedures requires securing the limbs with the animal in a prone position. Ligatures can be used to extend the limbs by attaching them to the corners of the surgical platform. Care must be taken to prevent overextension of the joints and to prevent compromise of circulation in the extremities.

To prepare the surgical site, the hair must be removed either by shaving or with depilatory cream. The hair removal should be conservative, removing only enough to provide a large enough surgical field to accommodate the incision and suturing. Excessive hair removal will intensify hypothermia (Bernal et al., 2009). The site is then prepared with a surgical scrub solution. The goal of the surgical scrub solution is to substantially reduce microbes present on the skin. Two commonly used scrub solutions are chlorhexidine (Nolvasan Scrub and Hibiclens) that are effective against bacteria and viruses even in the presence of organic matter and iodophores (Betadine Scrub, Prepodyne, Wescodyne , and Duraprep) that are not as efficacious in the presence of organic material. Rinses used between the scrubs are either sterile water or alcohol. Alcohol-based solutions containing 60–95% alcohol have great antimicrobial actions through the denaturing of proteins; however, alcohol can be a strong skin irritant, as well as serving to draw body heat away from the animal through evaporation. Sterile water is effective in rinsing the area, yet it does not have any antimicrobial properties.

8.5.2 DRAPING THE PATIENT

Surgical drapes are used to prevent contamination of suture material and to maintain a sterile field at the surgical site. Drapes are placed on the animal once it is prepared and positioned on the surgical platform. There are three types of draping materials: cloth reusable drapes, paper disposable drapes, and disposable plastic adhesive drapes. The use of disposable drapes allows for the surgeon to cut an opening or fenestration in the drape to any size or shape. A cloth drape will have a hole in it that cannot be altered. Although adhesive drapes are available as clear and opaque, for rodent surgery, the clear drapes are preferred as they allow for direct visualization of the patient. There are combination drapes made of both plastic adhesive and paper that allow the plastic to be placed directly over the surgical site and the paper to define the sterile field more clearly. Sterilized commercially available polyethylene wrap has been used for this purpose in rodent surgeries (Eakin et al., 2015). Not only is it cost efficient, but it also conserves the body heat of the animal, allows for direct visualization of the patient, and creates a moisture barrier around the animal. However, it must be used with caution as it can constrict movement for breathing when the wrap is placed around the patient too tightly.

8.5.3 SURGEON PREPARATION

For rodent survival surgeries (those in which the animal is allowed to recover from anesthesia), the surgeon should wear either a sterile surgical gown or a clean lab coat. Although it is not mandated that the surgeon wear a surgical mask, hair covering, or sterile gloves for rodent surgery, it is advised, especially when a body cavity is exposed. Prior to donning sterile gloves, the surgeon should thoroughly wash his or her hands and forearms with an antimicrobial soap and dry them with a sterilized towel. A sterile surgical gown and gloves are packaged such that the surgeon can put them on while maintaining sterility.

8.5.4 INTRAOPERATIVE CONSIDERATIONS

All instruments used during a survival surgical procedure must be sterilized. Experimental protocols may require the same surgical procedures to be performed on numerous animals. These are often done sequentially to optimize the surgeon's time. When serial surgeries are performed, the use of cold sterilants and hot bead sterilization can be employed to sterilize the instruments between patients. This facilitates the use of minimal instrumentation while still maintaining aseptic conditions.

The most common cause of anesthesia-related deaths in rodents is hypothermia. Because rodents have a high surface area to body mass ratio, they are prone to higher loss of body heat while being anesthetized because the anesthesia renders them incapable of shivering as a way to maintain normal body temperature. The best method to combat the heat loss is by supplying supplemental heat, such as by the use of heating pads or circulating water pads (Brauer et al., 2004). Although monitoring the animal's body temperature is also recommended, it is difficult with such small animals, as most thermometers are designed for larger animals. Rectal probes designed specifically for mice and rats are commercially available and their use is encouraged.

8.5.5 POSTOPERATIVE CARE

Once the surgical procedure is completed, the animal must be closely observed during the recovery phase. The animal should be recovered in a clean, well-ventilated recovery cage. Supplemental heat is provided until the animal is ambulatory. The animal is placed in lateral recumbency with the head and neck placed in a slightly extended position to prevent obstruction of the airway. Signs of postsurgical complications include the following: prolonged recovery, excitement, difficulty in breathing including rapid or shallow breaths, bleeding from the surgical site, pale or cyanotic mucous membranes, weak and rapid pulse, and coldness of body extremities. Incisions must be checked daily for indications of infection such as swelling, redness, exudate, or bleeding. Sutures must also be monitored and removed as designated by the experimental protocol, usually after 7–10 days.

Postoperative analgesics should be routinely used, particularly if an animal is exhibiting signs of pain. The rat (Sotocinal et al., 2011) and mouse (Langford et al., 2010) grimace scales that have been established to detect subtle signs of pain and distress in mice and rats can be used to assess the postsurgical patient. Orbital tightening, nose bulge, cheek bulge, ear position, and whisker change have been coded into a three-point scale: $0 =$ no pain, $1 =$ moderate pain, and $2 =$ severe pain levels. Once a technician has become familiar with the scale, rodent patients can be properly evaluated for postsurgical pain, and supplemental analgesics can be administered to the animal.

8.6 EUTHANASIA

Euthanasia is derived from Greek term meaning "good death." The term is used in biomedical research to describe ending the life of an animal in a manner that minimizes or eliminates pain and distress. Proper euthanasia techniques include a follow-up exam to confirm death, such as by determining the absence of a heartbeat. Monitoring respiration alone is not considered sufficient because there are euthanasia techniques in which the heartbeat may be maintained after visible respiration has ceased. The need to minimize fear and apprehension must be considered in determining the method of euthanasia. Distress vocalizations, fearful behavior, and release of certain odors or pheromones by a frightened animal may cause anxiety and apprehension in other animals. Therefore, animals should not be exposed to euthanasia of others.

The most common euthanasia method for mice and rats involves the use of carbon dioxide (CO_2) gas. As dictated by the American Veterinary Medical Association (AVMA, 2013), the use of CO_2 is acceptable with conditions that minimize aversion and distress. To minimize stress, the animals are left in their home cage, which is placed into a chamber in which CO_2 is gradually introduced at a displacement rate from 10% to 30% of the chamber volume per minute. As the oxygen is slowly replaced, the animals will lose consciousness prior to the perception of pain that is associated with nociceptor activation by carbonic acid. The flow is then maintained in the chamber once respiratory arrest has occurred to confirm that the animal is dead. An overdose of an inhalant anesthetic is also acceptable. Other methods are listed in the AVMA guidelines; however, some of these methods require scientific justification as they may cause pain or distress to the animal.

8.7 NECROPSY

Many studies include data collected from the research animal at multiple time points, including postmortem. The requirements for the necropsy are dependent on the nature of the research. If sequential sampling is to be done to remove many of the organs, it must be done as quickly as possible to avoid postmortem changes. The animal protocol should dictate the specific tissue/organ samples to be collected and the method of collection. Once the tissues and organs are removed from the body, they will need to be properly stored in labeled containers with the correct fixative solution, usually buffered formalin.

It is advisable to perform necropsies under a biological safety cabinet or on a downdraft necropsy table. Personnel performing or observing the necropsy should wear appropriate protective equipment, such as gloves, eye protection, respiratory protection, and a lab coat. A standard record form should be established to note observations made during the procedure. If a necropsy is performed in response to a lethal disease or other medical concern, observations should be made of the entire body, including the condition of the skin and fur. When the body cavity is exposed, observations should be made with the organs in place. The organs can then be removed for diagnostic testing such as histology, parasitology, and microbiology.

8.8 SUMMARY

The handling and performance of technical procedures on experimental animals must be done by fully trained and proficient personnel. Training materials should include the use of correct instrumentation, proper animal handling techniques, and a review of the regulations and guidelines that dictate the use of animals for experimentation. Whenever animals are used for an experiment, precaution must be taken to ensure that the procedures are done on the right animal, at the right dose and route, and with the right compounds. These can be referred to as the "4 Rights" of animal experimentation. Animal identifications should be confirmed whenever the animals are handled. Euthanasia should only be performed after the identification of the animal is verified.

REFERENCES

AALAS Learning Library, 2005a. Animal handling and restraint. In: Working with The Laboratory Mouse. https://www.aalaslearninglibrary.org/Pages/Courses/course.aspx? intCourseID=2451&intPageID=94993.

AALAS Learning Library, 2005b. Animal handling and restraint. In: Working with The Laboratory Rat. https://www.aalaslearninglibrary.org/Pages/Courses/CourseAllPages.aspx?intLessonID=27691.

Adeghe, A.J.-H., Cohen, J., 1986. A better method for terminal bleeding of mice. Lab. Anim. 20, 70−72.

American Veterinary Medical Association, 2013. AVMA Guidelines for the Euthanasia of Animals: 2013 Edition. https://www.avma.org/KB/Policies/Documents/euthanasia.pdf.

Balcombe, J.P., Barnard, N.D., Sandusky, C., 2004. Laboratory routines cause animal stress. Contemp. Top. Lab. Anim. Sci. 43, 42−51.

Bernal, J., Baldwin, M., Gleason, T., Kuhlman, S., Moore, G., Talcott, M., 2009. Guidelines for rodent survival surgery. J. Invest. Surg. 22, 445−451.

Brauer, A., Perl, T., Uyanik, Z., English, M.J.M., Weyland, W., Braun, U., 2004. Perioperative thermal insulation: minimal clinically important differences? Br. J. Anaesth. 92, 836−840.

Cesarovic, N., Nicholls, F., Rettich, A., Kronen, P., Ha, M., Jirkof, P., Arras, M., 2010. Isoflurane and sevoflurane provide equally effective anaesthesia in laboratory mice. Lab. Anim. 44, 329−336.

Diehl, K.-H., Hull, R., Morton, D., Pfister, R., Rabemampianina, Y., Smith, D., Vidal, J.-M., van de Vorstenbosch, C., 2001. A good practical guide to the administration of substances and removal of blood, including routes and volumes. J. Appl.Toxicol. 21, 15−23.

Eakin, K., Rowe, R.K., Lifshitz, J., 2015. Modeling fluid percussion injury relevance to human traumatic brain injury. In: Kobeissy, F.H. (Ed.), Brain Neurotrauma: Molecular, Neuropsychological, and Rehabilitation Aspects. Taylor & Francis Group, Boca Raton, pp. 259−274.

Hem, A., Smith, A.J., Solberg, P., 1998. Saphenous vein puncture for blood sampling of the mouse, rat, hamster, gerbil, Guinea pig, ferret and mink. Lab. Anim. 32, 364−368.

Institute for the Laboratory Animal Research, 2011. Guide for the Care and Use of Laboratory Animals, eighth ed. National Academies Press, Washington, DC.

Joslin, O.T., 2009. Blood collection techniques in exotic small animals. J. Exot. Pet Med. 18, 117−139.

Langford, D.J., Bailey, A.L., Chanda, M.L., Clarke, S.F., Drummond, T.E., Echols, S., Glick, S., Ingrao, J., Klassen-Ross, T., LaCroix-Fralish, M.L., Matsumiya, L., Sorge, R.E., Sotocinal, S.G., Tabaka, J.M., Wong, D., van den Maagdenberg, A., Ferrari, M.D., Craig, K.D., Mogil, J.S., 2010. Coding of facial expressions of pain in the laboratory mouse. Nat. Methods 7, 447−449.

Machholz, E., Mulder, G., Ruiz, C., Corning, B.F., Pritchett-Corning, K.R., 2012. Manual restraint and common compound administration routes in mice and rats. J. Vis. Exp. 67, e2771. http://dx.doi.org/10.3791/2771.

NIH.gov, 2010. Guidelines for the Survival Bleeding of Mice and Rats. acu.od.nih.gov/ARAC/documents/Rodent_Bleeding.pdf.

Robinson, V., Morton, D.B., Anderson, D., Carver, J.F.A., Francis, R.J., Hubrecht, R., Jenkins, E., Mathers, K.E., Raymond, R., Rosewell, I., Wallace, J., Wells, D.J., 2003. Refinement and reduction in production of genetically modified mice. Lab. Anim. 37, S1−S50.

Sirois, M., 2016. Laboratory Animal Medicine: Principles and Procedures, second ed. Mosby, St. Louis.

Smith, W., 1993. Responses of laboratory animals to some injectable anaethetics. Lab. Anim. 27, 30−39.

Sotocinal, S.G., Sorge, R.E., Zaloum, A., Tuttle, A.H., Martin, L.M., Wieskopf, J.S., Mapplebeck, J.C.S., Wei, P., Zhan, S., Zhang, S., McDougall, J.J., King, O.D.,

Mogil, J.S., 2011. The rat grimace scale: a partially automated method for quantifying pain in the laboratory rat via facial expressions. Mol. Pain 7. http://mpx.sagepub.com/content/7/1744-8069-7-55.full.pdf+html.

Tsukamoto, A., Serizawa, K., Sato, R., Yamazaki, J., Inomata, T., 2015. Vital signs during injectable and inhalant anesthesia in mice. Exp. Anim. 64, 57−64.

Turner, P.V., Brabb, T., Pekow, C., Vasbinder, M.A., 2011a. Administration of substances to laboratory animals: routes of administration and factors to consider. J. Am. Soc. Lab. Anim. Sci. 50, 600−613.

Turner, P.V., Pekow, C., Vasbinder, M.A., Brabb, T., 2011b. Administration of substances to laboratory animals: equipment and considerations, vehicle selection, and solution preparation. J. Am. Soc. Lab. Anim. Sci. 50, 614−627.

CHAPTER

Considerations for Use of Vertebrates in Field Studies

9

M.J. Cramer

University of Notre Dame, Notre Dame, IN, United States

CHAPTER OUTLINE

9.1 REASONS FOR CONDUCTING FIELD RESEARCH

There are several situations where conducting research with animals in the field is necessary. Historically, most field research with animals has been associated with the documentation of animal populations, but more recently, the ability to address questions of physiology and behavior has become more commonplace. Technological advances have allowed more remote studies to be conducted on free-living animals.

Use of wild-caught individuals over lab-reared animals has several benefits and drawbacks. The main benefit of wild-caught animals concerns the direct interpretation of physiology and behavior in a realistic evolutionary context. What is sacrificed is the high level of control of any potentially confounding variables. For example, when using wild-caught mice to study mate choice, assumptions must be made about the level of experience the subjects have with each other. Even though field protocols can be used to minimize the effects of these biases, for instance by only using mice from different trapping grids (Cramer and Cameron, 2007), nothing can replace the certainty achieved by using mice raised in the laboratory. There has been much discussion in the literature about the use of wild-caught over lab-reared animals in behavioral and physiological experiments. Given the trade-off between experimental control and evolutionary or ecological context, studies of behavior and physiology should include both field and laboratory components (Calisi and Bentley, 2009).

Regardless of the ultimate goals of these studies, field research on vertebrates requires the capture and handling of live animals. In most cases, animal survival is essential. Moreover, most field situations do not allow for timely veterinary care, so attention must be taken with the capture and handling of animals. Most zoologically based professional societies have published guidelines for the capture and handling of different animals (Table 9.1). Whenever working with wildlife, animal

Table 9.1 List of Published Guidelines for the Capture and Handling of Different Animal Taxa Published by Professional Societies

Taxon	Society	References
Fish	American Fisheries Society	UFR Committee (2014)
Amphibians and reptiles	American Society of Ichthyologists and Herpetologists	HACC (2004)
Birds	The Ornithological Council	Fair et al. (2010)
Mammals	American Society of Mammalogists	Sikes and Animal Care and Use Committee of the American Society of Mammalogists (2016)
Animals	Animal Behavior Society	Animal Care Committee (2012)

Online copies of many of these resources are available at the United States Department of Agriculture website, in the Animal Welfare Information Center (www.nal.usda.gov/awic).

welfare should be considered but is often overlooked (Cattet, 2013). Selection of techniques for the capture, marking, handling, and any additional physiological or behavioral measures should minimize animal disturbance. The more disruptive and invasive the technique, the more likely individuals or populations will be negatively affected, which could lead to erroneous conclusions, especially concerning animal behavior or population estimation.

One way in which field investigations dealing with live animals differs from those of laboratory studies is in the potential scope of effects on the animals themselves. Unlike many laboratory studies in which the study organism are lab-bred, field studies have the additional concern for the wild populations of subjects. Therefore, field researchers have potential effects not solely on the animals with which they are dealing but the populations from which those animals originate (Curzer et al., 2013). This is especially true for studies of threatened or endangered species, and there is increased ethical responsibility concerning the proper use and care of these animals (Minteer and Collins, 2005). Moreover, field researchers need to be cognizant of the potential long-lasting effects on wild population of environmental manipulations (Cuthill, 1991; Farnsworth and Rosovsky, 1993).

9.2 SOURCES FOR SPECIFIC TECHNIQUE INFORMATION

Apart from this chapter, which provides a general overview of working in the field with vertebrates, there are many additional sources that provide specific details about techniques used for vertebrates. Including the guidelines from professional societies (Table 9.1), manuals have been published for amphibians (Heyer et al., 1994; Dodd, 2009), mammals (Wilson et al., 1996), fish (Zale et al., 2012), bats (Kunz and Parsons, 2009), birds (Bub, 1995), and wildlife in general (Braun, 2005). These sources should be referenced for more details on working with specific vertebrate groups.

9.3 INITIAL STEPS FOR UNDERTAKING FIELD RESEARCH INVOLVING VERTEBRATES

9.3.1 OBTAINING PERMISSION FROM LANDOWNERS

Successful field studies with animals require planning. In addition to the acquisition of equipment and supplies for the project, researchers usually need to obtain permission at several levels to conduct field research. First, permission to conduct any field study must be obtained from the person or organization that owns the land where the study will take place. Getting this permission can sometimes determine where the study is to take place. Private landowners need to be contacted in advance, and responses to scientists wanting to conduct research on their land range from friendly to paranoid. Permission to conduct fieldwork on state and federal lands is usually fairly easy to obtain and generally consists of applying for a permit.

However, as with other bureaucracies, researchers need to factor in up to 6 months for the proper forms to be filed. Private research facilities also have their own protocols for researchers to obtain permission to work there. In most cases, researchers file a research plan with the facility outlining project goals, number of animals used, potential long-term effects, and other concerns specific to the facility.

9.3.2 INSTITUTIONAL ANIMAL CARE AND USE COMMITTEE APPROVAL OF ACTIVITIES

As with any research concerning live animals, a protocol must be submitted and approved by an Institutional Animal Care and Use Committee (IACUC). More information concerning this process can be found in the Suckow and Lamberti Chapter 4 of this book, but there are some considerations specific to field research.

There are many inherent differences between biomedical and wildlife studies (Sikes and Paul, 2013). While historically, IACUCs and the guidelines used to assess proper care and use of animals were developed with biomedical research in mind, recently there has been a call for IACUCs to consider some of the issues regarding field research (Sikes et al., 2012). The most reasonable solution is to accept published guidelines from scientific professional societies as the standard for the IACUC approval. In other words, protocols that have been vetted by scientific societies and are accepted as common practice are often regarded as adequate for IACUC approval. Moreover, members of IACUCs should recognize the oversight of federal and state agencies granting permits that govern the use of animals in field research (Paul and Sikes, 2013).

One major difference between field- and laboratory-based protocols is the acquisition and number of animals used. The acquisition of animals is more complicated in field-based studies because animals must be captured from wild populations, which carries two problems that biomedical researchers do not have to address: (1) the capture of nontarget species and (2) the number of animals used for a given protocol. Capture of animals from wild populations incorporates a certain amount of uncertainty, which must be addressed in IACUC protocols. First, no matter which trapping technique is used, there is always a high probability of capturing nontarget species. Thus IACUC protocols need to include a list of every possible species that could be inadvertently captured. For example, even though the author's research is concentrated on a single species, the woodland deer mouse, small mammal traps are used to obtain subjects; thus it is essential to include a list of every small mammal species that occurs at the study sites, including chipmunks, flying squirrels, jumping mice, voles, shrews, and other mouse species. Second, it is impossible to accurately predict precisely how many animals will be captured prior to trapping. However, the number of animals to be used is an important consideration for most IACUCs as using more animals than necessary could lead to a regulatory compliance concern. Therefore it is best to provide an estimate of the number of animals to be used, based on the number of traps and an estimate of the capture rates (from earlier trapping efforts or gleaned from the literature). It is also helpful to include these estimates in the protocol, so the members of the committee understand that there is some uncertainty in the numbers.

In addition, because animals are obtained from wild populations, researchers have an ethical responsibility not to oversample or cause any lasting effects on these populations or their habitat (Curzer et al., 2013). In particular, for studies of rare or endangered species, with small population sizes, researchers should be careful not to affect population growth rates. This includes removing individuals for extended periods of time, which could have negative effects on territory defense or limit mating opportunities for those individuals. Use of animals from wild populations should be as limited in time as possible to avoid these potential effects.

9.3.3 PERMITS

For all vertebrate species, state and federal permits need to be obtained. Some permits are relatively easy to obtain, and the regulations vary from state to state. In most cases, for vertebrates except for birds, permitting is handled by the Department of Natural Resources for that particular state. An important consideration for permitting is the use of state-listed species, which are species that are considered endangered or at risk. Most states have a scientific permit that allows use of most vertebrates for scientific study or education. Species lists and the number of individuals used are usually reported to the state annually.

Use of some vertebrates requires federal permits, which can be harder to obtain. Birds are an excellent example. To be able to handle birds, permittees must be trained in bird-banding techniques, a process that requires working with another scientist who already has a bird-banding permit through an intensive program which can take several weeks to complete. The reason behind this restriction is that many passerine species are awarded federal protection, as they are neotropical migrants.

9.4 OBSERVATIONAL TECHNIQUES

For some vertebrate groups, it is preferable to use low-impact observational techniques as opposed to capturing individuals. One of the major advantages to using these techniques lies in the minimal effect they have on animal subjects. At the very least, observational techniques may have short-lived effects on immediate behavior. Another strength of these techniques is the ease with which they may be implemented; they require little training short of identification of species and have been used successfully in citizen science projects (Chase and Levine, 2016). To obtain density data, the area sampled must be taken into consideration; however, estimates of density are not as accurate as those obtained by capturing and marking animals.

Regardless, many researchers use transect methods to delineate area sampled. Transects allow for the collection of data over larger distances, including various habitat types or gradients. There are two basic approaches to using transects, line transect sampling and strip transect sampling (Burnham and Anderson, 1984). Line transect sampling involves following a predetermined transect of given length and counting every individual as well as their distance from the center of the transect. Distance measurements help eliminate the bias introduced by unobserved individuals because it is assumed that detectability decreases with the distance from the center of the transect. Strip transect sampling differs in that all individuals within a given distance from the center of the transect are counted. An important consideration in the use of transect is randomization. Transects should be relatively short to allow for replication. Moreover, to satisfy statistical assumptions, the location of transects should be randomly determined as much as possible.

9.4.1 POINT COUNT SURVEYS

These are usually used for studies of birds. The premise is simple: investigators travel a predetermined route, stopping at regular intervals to count all birds that are seen or heard within a given time period. This technique has been used successfully for a long time, and the most well-known application is the Christmas Bird Count and the Breeding Bird Survey. Obviously, the effectiveness of this technique is dependent on the skills of the observers, and generally some training should occur prior to sampling to make sure certain routes are not more accurately sampled than others.

9.4.2 CALL SURVEYS

This technique is generally used for anurans and yields the most information during the breeding season. Call surveys are similar to point count surveys, except that during call surveys, investigators generally will go to a specified place, such as a lake or pond, where amphibian residence is expected. This technique is useful for species that occur in forested habitats and are primarily nocturnal or crepuscular. In call surveys, researchers arrive at the breeding location, allow a set amount of time for animals to settle, and start the survey. At its simplest inception, the species identity of calling frogs and toads is recorded, along with a measure of call intensity (usually ranging from 0 to 5). Call intensity is somewhat subjective and should be used to estimate relative abundance. Calls can also be used with a transect design to locate nocturnal or hard-to-see species (audio strip transects; Zimmerman, 1994).

9.4.3 VISUAL ENCOUNTER SURVEYS

Visual encounter surveys are a simple technique useful for diurnal species that are easily identified by sight. For this technique, researchers walk around in a specified area and record all individuals seen. Sampling designs vary and can range from a random walk to more spatially explicit transects or quadrats. Intensity of search can also vary depending on the goals of the study. The least intensive surveys only count individuals encountered on the surface; these surveys take the least amount of time but tend to miss rarer species. For highly intensive surveys, all surface objects are overturned and torn apart to find individuals; leaf litter is raked to find concealed individuals. Highly intensive surveys obviously do the most damage to the immediate environment, but they are most likely to find all species in the study area. For example, highly active amphibians, such as frogs, can easily be observed with a less intense survey, but other species, such as salamanders, are missed. Intermediate-intensity searches can be used to increase the number of species encountered without undue damage to natural areas. For these searches, animals encountered on substrate surfaces are counted, along with those found by turning over rocks and logs. Objects turned over are replaced to their original location so as to not have any lasting effects on the microhabitats.

9.4.4 CAMERA TRAPS

One shortcoming with many observational techniques involves the inherent bias due to the presence of the observer. By moving through the habitat, observers may frighten shyer species, thereby making them less likely to be included in data collection. Other species may actively flee, making their identification difficult. One way to reduce this bias is to observe animals remotely with the use of camera traps (also referred to as trail cameras). Camera traps are placed in particular habitats and take pictures of animals when they cross an infrared beam, which triggers the camera to fire. Researchers periodically check the cameras to remove media containing pictures or videos and replace batteries if necessary. With the development of digital photography, use of camera traps has increased greatly. Although historically used for monitoring ungulates and carnivores, camera trap technology has advanced to include the capability of photographing smaller-bodied animals, including armadillos and rodents (Ahumada et al., 2013). In addition, modern camera traps can record short videos to collect behavioral data as well (Lobo et al., 2013; Nations and Olson, 2015; Flagel et al., 2016). Considerations when using camera traps for density estimation are similar to other techniques and include observability and detection probability, ability to read marks or identify individuals, and estimation of area sampled. These issues need to be explicitly addressed and solutions should be standardized (Burton et al., 2015). In response to their increased use in wildlife and ecology studies, several guides have been published to describe the use of camera traps in field studies (O'Connell et al., 2011; Meek et al., 2014).

9.5 CAPTURING ANIMALS

Capturing animals in a manner that is not harmful to individuals or populations is essential in any field study of vertebrates. This section will focus on techniques utilizing traps and nets to restrain animals; Schemnitz (2005) provides species-specific information for most types of wildlife. Most methods used to capture animals have this in mind, but the best way to ameliorate the potential negative effects of capture on individuals is checking traps and nets regularly. The longer an animal is contained in a trap or net, the more likely there will be a catastrophic event, including predation, injury, or death due to exposure. Moreover, researchers are responsible for removing traps at the conclusion of their study, to ensure that animals are not unknowingly trapped and die of exposure.

When capturing animals, the safety of the animals is the researcher's primary concern. There is nothing more pointless and wasteful than neglecting to check open traps and nets, which could lead to high levels of animal death. The main priority when traps are open is to reduce the possibility of animal death or injury by checking them in a timely manner.

The placement of traps and nets is essential because location can seriously affect capture rates. Particularly with nets, care must be taken to moderate capture rates so that animals are not left struggling in the net while others are being removed. An important consideration when placing traps is exposure. Traps should be placed so that captured animals are not left out in the open. This is particularly a problem with aluminum traps such as Sherman traps because captured animals are likely to overheat and go into torpor if the trap is left in direct sunlight for extended periods. Even with cage traps, captured animals without cover are more susceptible to predation. Providing cover for captured animals either by covering the traps with leaves or providing a nest box within the trap (e.g., for larger Tomahawk traps) can significantly increase trap survival.

9.5.1 ELECTROFISHING

One technique used specifically for the capture of fish is electroshocking. An electric current is used to temporarily stun fish for capture and processing. It can be used in a variety of habitats that require use of specialized equipment. For stream sampling of smaller fish, a backpack electroshocker is employed. Backpack electroshockers are also useful for catching small young-of-year fish along shoreline habitat. This is sometimes preferable to nets, which may snag and damage complex habitat features, such as coarse woody debris. For larger fish in lacustrine habitats, a shock boat may be used.

Different types of electrical current can be used to catch fish. Direct current (DC) flows in a single direction (cathode to anode) whereas alternating current (AC) changes direction. Use of DC electrofishing forces fish to swim in the direction of the current. There is the potential for catching different species depending on the type of current used; Porreca et al. (2013) reported higher numbers of fish captured using AC fishing, which was driven by higher captures of sunfish and catfish. Captures of largemouth bass and crappie were equivalent. However, use of AC may be more damaging to fish and lead to higher mortality (Snyder, 2003; Awata et al., 2013).

There are five basic responses of fish exposed to electrical current (in increasing severity): (1) avoidance; (2) forced swimming (electrotaxis); (3) muscle contraction (electrotetanus); (4) muscle relaxation (electronarcosis); and (5) death (Meador et al., 1993). Obviously, electronarcosis is the desired result of electrofishing; care must be taken to prevent fish death. The higher the frequency of pulses per second (pps) used, the more likely the fish will be killed. Meador et al. (1993) suggest a frequency between 30 and 60 pps for most species. Consideration of water conductivity is also important and should be measured prior to electrofishing to determine the ideal output voltage for optimum fishing efficiency. Low-conductivity water (<20 μS/cm) is resistant to electrical currents and would require higher voltage to properly shock fish. The size and strength of the generated electrical field should be tested prior to electrofishing.

Safety is a major concern during electrofishing. Crew members should be wearing proper safety equipment to prevent electric shock, including rubber boots (or waders) and rubber gloves. In addition, those collecting fish should have polarized glasses to be able to see fish as they are shocked. As with any field exercise, researchers should be alert and aware of their surroundings to prevent personal injury. Safety is also a concern for fish being sampled: it is relatively easy to injure fish unintentionally during electrofishing. Sustained exposure to an electrical current will lead to fish death by asphyxiation. Injuries related to excessive muscle twitching such as muscular hemorrhages and spinal injuries can occur at high pulse frequencies. Researchers should be alert to these problems, as they can be difficult to observe in living fish (Snyder, 2003). Moreover, stunned fish should be removed from the electrical current quickly and placed in a holding box or live well for processing. To reduce mortality, live wells should be properly aerated. Finally, fish should be released downstream to prevent recapture.

9.5.2 NETS

A wide variety of nets are used to capture fish. Use of nets to capture fish can be broadly divided into passive and active techniques. Passive techniques involve leaving nets unattended for a set amount of time to collect animals. In contrast, researchers using active techniques will move nets, mainly through trawling, to capture individuals.

Passive methods capture individuals through entanglement or entrapment (Hubert et al., 2012); the main difference lies in how fish interact with the net. Entanglement nets, such as gill nets, hold animals in place by physically wrapping around spines, gills, fins, or other morphological structures. This can lead to an increase in mortality, especially if entanglement occurs on the gills. On the other hand, entrapment nets, such as hoop or fyke nets, keep fish contained in a small area. Entrapment nets have a lower mortality rate than entanglement nets (Hubert et al., 2012) because fish are not physically trapped within the net material itself. One concern with use of passive techniques is bycatch or the capture of additional species besides the species of interest. This is not restricted to nontarget fish species but includes other aquatic and semiaquatic vertebrates as well, such as turtles, beavers, and muskrats. In areas where these species may be captured, active techniques should be used, or air space must be left at the top of the net to reduce the probability of drowning these creatures.

Although more labor-intensive, active techniques allow for a greater capture rate in a shorter period of time (Hayes et al., 2012). In active techniques, nets are monitored and controlled by the researcher and are useful for catching smaller species, especially those that occur in streams. One commonly used net is a seine, which is a net stretched between two poles. The net is placed downstream and held in place by two people. Meanwhile others disturb the substrate upstream and drive fish, and other stream denizens, into the net, where they can be taken out and processed. Seines can also be used in conjunction with electrofishing to increase catch (Meador, 2012). In more expansive habitats, such as lakes, active techniques include trawling, in which a net is dragged behind a moving boat, and pelagic fish are swept up. Finally, fish can also be sampled by angling, but the catch per unit effort is highly variable and depends on the skill of the angler.

Nets are also used to capture volant vertebrates. Birds and bats are captured in flight using mist nets, which are constructed of fine thread so animals in flight do not see them until they are entangled. Species that forage for tiny insects, such as swallows and some bats, can see and avoid mist nets. Nets are placed in foraging areas, usually against a cluttered background, which makes it more difficult for animals to see the net. Nets should be monitored closely, as struggling animals are harder to remove, and in the case of bats, they will eventually chew their way through the net. Additional information can be found in Kunz et al. (1996) and Kunz and Parsons (2009).

9.5.3 TRAPS

Many vertebrate species are captured using traps. Most traps work best in the absence of the trapper and utilize the natural movements of animals for capture. As a general rule, simpler traps are more effective. The more complicated the trap mechanism becomes, the more likely it will malfunction, thereby reducing trap effectiveness. Understanding the natural history of the target species is a key for increasing trap efficiency, especially as it relates to bait choice and trap placement. For example, attempts to trap carnivores, which are reliant on the sense of smell for hunting, will be unsuccessful if the trapper is careless about leaving his or her own scent on or near the trap.

Traps used in fisheries are referred to as pot gears, which are traps usually left and baited in fish habitat. As with any type of trap, care must be taken to ensure that all traps are removed at the completion of the study to avoid "ghost fishing" (Guillory, 1993), or catching individuals without removing them from the trap, leading to high mortality. Lost pot gears are particularly susceptible to this problem because expired individuals can serve as bait to capture additional fish.

Pitfall traps are commonly used to catch mobile amphibians and reptiles but are also effective for small mammals such as shrews, which due to their small size are more difficult to catch in box traps. Pitfall arrays are time consuming to implement but will catch any mobile species in the area. These traps are simply containers placed underground with their openings flush with the ground surface. The bottom of the containers should be perforated to allow water to pass through freely, so captured animals will not drown if heavy rains occur. Some researchers working with amphibians or other species prone to dehydration will include damp sponges inside the traps. Capture efficiency can be improved by placing traps in areas with increased activity (e.g., runways or near optimal habitat) as well as by connecting individual traps with drift fencing to funnel animals toward the traps. The depth of the traps is dependent on the target species. Pitfalls are commonly used to capture anurans, so they should be deep enough so they cannot simply jump out. Escapes can be prevented by placing a plastic collar around the trap opening (Corn and Bury, 1990), although the effectiveness of rims can be species dependent.

9.5.3.1 Box Traps

Traps used to capture most mammals are box traps. Box traps vary in size and construction, but a few trap types have become the "industry standard." Sherman traps are generally used to study small mammal species, especially shrews and rodents, because they are simple and easy to use. They come in a variety of sizes: smaller ones designed for shrews (trap dimensions: $2 \times 2.5 \times 6.5$ in.) and larger traps for squirrel-sized animals (trap dimensions: $15 \times 5 \times 5$ in.). The trap mechanism is relatively simple. The trap is set with a treadle holding the front door open. The animal enters the trap, moves to the back, and their weight moves the part of the treadle holding the door open, closing the door behind them. Sherman traps are constructed from aluminum and certain models are collapsible, making them easy to transport to the field. However, because they are made of aluminum, captured individuals can die easily from exposure if left in the trap too long. For this reason, traps need to be checked regularly. In open areas with lots of sun, researchers sometimes construct small covers to keep traps shaded. Another way to avoid trap deaths is to complete trap checks early in the morning, before ambient temperatures get too high. In addition, traps can be provisioned with extra bait and nesting material such as cotton to ensure overnight survival on cold nights. Cold overnight temperatures lead to increased trap mortalities, so careful deliberation should be taken before setting traps on nights where the predicted temperature is too low. This limit is dependent on the target species. For example, a researcher who included cotton nesting material in traps set for forest mice would be able to trap at lower temperatures, but a researcher trapping kangaroo rats could not because this species does not use nesting material.

Although Sherman traps come in a variety of sizes, the box traps typically used to catch larger mammals are Tomahawk traps. These are used to catch chipmunk-sized (trap dimensions: $18 \times 5 \times 5$ in.) to coyote-sized (trap dimensions: $72 \times 20 \times 26$ in.) animals. These traps consist of wire mesh frames with a simple trap mechanism: bait is placed behind a treadle which is attached by a metal arm to a hook holding the trap door open. Unlike Sherman traps, Tomahawk traps have a cage design, so although captured animals are somewhat exposed, thermoregulation is not as much of a concern compared to Sherman traps. However, there is a greater incidence of captured animals injuring themselves as they are sometimes able to fit extremities between wires of the trap. The author has seen some chipmunks rub the skin off their snouts trying to escape from a trap and has witnessed marauding raccoons harassing captured rodents in Tomahawk traps. These issues can be ameliorated by either placing a small nest box inside the trap or placing a trap cover over the trap. Trap covers can be purchased or easily constructed from a variety of materials, such as pieces of tarp or shadecloth. Covers not only reduce stress for captured animals, but also provide some relief from high temperatures and precipitation. In addition, they increase capture success by reducing opportunities for bait stealing; interested animals are forced to access the bait solely by using the front opening of the trap.

Bait is used for box traps for two reasons. First, bait is used as an attractant to increase capture rates. Most mammals forage using smell, so use of odorant baits is more effective at increasing trap success. Attractants are generally used for larger species with large home ranges, especially carnivores. For small mammal trapping, fragrant baits are more likely to attract predators, such as raccoons, which may destroy traps to get to the bait or any trapped animals, thus endangering any small mammals already in the traps. The second reason for including bait in a trap is to provide sustenance for captured animals until the next trap check. This is more important for small mammals, which may not have the energy reserves to survive an extended period without food. For herbivorous mammals captured during the day, the fruit is excellent bait because it provides both energy and water for captured animals.

At the conclusion of trapping, all traps should be accounted for and then cleaned for three main reasons. First, during their time in the trap, most animals expel urine and feces, which can increase the transmission of diseases and parasites to future captures if the trap is not cleaned properly. Second, cleaning traps ensures that they continue to function properly. Unused bait and fecal material can accumulate under the treadle, which interferes with the sensitivity of the trap, and makes it harder to set properly. Third, many mammals are sensitive to particular odors, so the smell of previous occupants may affect trap success. To eliminate both potential disease agents and residual odors a 10% bleach solution is commonly used to clean traps.

9.6 HANDLING ANIMALS
9.6.1 PHYSICAL RESTRAINT

The most important consideration when handling animals is the health and safety of both the handler and the animal being handled. This can be achieved with careful selection of personal protection equipment and experience. In addition, there are a variety of instruments that can be used to control and subdue larger animals.

For smaller animals, it helps to transfer the animal to a smaller container, which allows the handler to locate and restrict animal movement. For example, when trapping mice, the animal can be transferred from a trap to a clear plastic bag prior to handling. This allows the handler to see what the animal is doing, and monitor its well-being. By folding the bag, one can control the location and movement of the mouse. Soft flexible nets are preferable for small fish species as they minimize loss of scales and mucous (Harmon, 2009). For larger mammals such as squirrels a handling cone can be easily constructed of canvas (Koprowski, 2002). The cone allows access to different parts of the animals, while simultaneously keeping the face covered, thus reducing the stress of being handled.

A natural reaction of an animal to being handled is to attempt to escape. In some cases, its eyes and face can be covered, which tends to reduce the stress of handling (Rudran and Kunz, 1996). Control is a key to successful handling. There is a delicate balance between holding an animal too firmly, which could result in animal injury or death, and holding an animal too loosely, which could result in escape or handler injury. Practice is the best way to learn this balance. Any animal with a mouth can bite, and biting should be anticipated. In some cases, biting can be incorporated into the handling process. Having control over where an animal bites can reduce any surprise attacks that could lead to escape. In the case of chipmunks, allowing the animal to bite a gloved finger allows the handler to secure the back of its head with the thumb. This hold provides the most control over the animal and reduces struggling. With particularly feisty species, securing the animal's body as well as the head is necessary. The body can be secured by holding the base of the tail or the hips. Use of the tail for handling can be dangerous, as this can lead to injury in some species. Thus tail-holding should be used to secure the animal to prevent escape, and animals should not be held up by their tails. The handler should remain calm and attentive.

Amphibians require special consideration. Because their skin is a respiratory organ, great care should be taken when handling them. If gloves are not used, the hands should be clear of all potential chemical contaminants, especially insect repellent. Another consideration for many amphibians is desiccation. Water, preferably from the habitat, can be poured over the animal if the skin starts to dry out (Fellers et al., 1994). If temporarily being held in a secondary container a small amount of water from the habitat or some wet vegetation should be placed in the bag to allow additional moisture.

9.6.2 CHEMICAL RESTRAINT

Many vertebrates can be more easily handled after use of drugs to immobilize or sedate them. Considerations for the use of chemical restraint include (1) which particular drug to use, (2) the delivery system to get the drug into the animal's bloodstream, and (3) how to allow the subject to recover property after being restrained.

There are a variety of delivery systems, mainly determined by the temperament and size of the animal being drugged. For small mammals, an inhalant (gas) anesthetic, such as isoflurane, can be used in a chamber that restricts movement of the animal. For larger mammals such as raccoons, animals can be drugged with a syringe while still inside the trap, simplifying the handling process. It is essential that the animal being drugged is in a relatively quiet area, free of loud noise, and distractions.

Many factors need to be considered when choosing a drug for restraint, including the particular therapeutic index for the drug (TI = lethal dose/effective dose), the induction period, if the drug has an antagonist, and the solution stability of the drug (Fowler, 1995). Drugs with a high TI allow for a greater margin of error with respect to overdosing an animal. Induction period, or the amount of time it takes for the drug to take effect and immobilize the animal, is critical for wild species. Drugs with a shorter induction period are preferred, especially for highly mobile animals which may not be contained. A partially drugged animal is more likely to injure itself or become a victim of predation should it escape the researcher before the drug takes effect. Antagonists are agents that quickly reverse the effects of a particular chemical agent. Thus use of drugs having available antagonists is preferred as they shorten the recovery period. Finally, for field situations the solution stability of the drug is important. Drugs that need to be kept at cool temperatures are not always useful in the field, where refrigeration may not be readily available. Common drugs used for restraint include isoflurane, ketamine, telazol, and MS 222 (tricane methanosulfanate or ethyl M-aminobenzoate methanosulfanate). It is important that captured animals do not develop a preference for the drug being used. For example, raccoons have been shown to demonstrate a preference for telazol, leading to increased recapture rates (Gehrt et al., 2001).

Another consideration is the delivery systems for anesthetics. The two main avenues for the delivery of drugs into a captured animal's system is injection into the bloodstream or through inhalation. Delivery methods vary based on the species being anesthetized. Most small mammals are easily anesthetized by inhaled agents, whereas fish and amphibians can be anesthetized with MS 222 (tricane methanosulfanate or ethyl M-aminobenzoate methanosulfanate) absorbed through the gills in fish and larval amphibians or through the skin in adult amphibians. A 30% ethanol solution can also be used to restrain adult amphibians but may not be effective for all species (Fellers et al., 1994). Drugs are applied in solution to a small container housing the animal. For use of MS 222 with adult amphibians, the solution used should be at a neutral pH to avoid incidental skin damage, and the dosage should be relatively small (0.03−0.05%; Fellers et al., 1994).

Inhaled anesthetics are a bit more difficult to use in the field. In most field situations, it is not feasible to transport heavy equipment, such as gas canisters and a precision vaporizer, for the delivery of drugs. Mammals can be anesthetized using gas anesthetics such as isoflurane or halothane through the use of a chamber or a face mask (Parker et al., 2008). The chamber method works well for mammals in cage traps, which have a fairly constant air flow into and out of the trap and allow for a clear line of sight to monitor the animal's reaction to the drug. The entire trap is placed into an air-tight chamber, and the animal is closely monitored until it is unresponsive. Although this method is preferable because it reduces the stress experienced during handling, it is not as efficient because the drug needs to permeate the entire air space in the chamber before it can effectively neutralize the animal. Inhalant anesthetics can also be administered via equipment that requires use of gas canisters (Lewis, 2004) or by applying the drug to cotton or some other material and placing it in the trap with the animal (Parker et al., 2008). Caution should always be used when working with gas anesthetics to minimize the accidental exposure of personnel to waste gases.

Small mammals or birds in the hand can be effectively anesthetized using a face mask or nose cone, which can be constructed inexpensively by converting a centrifuge tube (Parker et al., 2008). Cotton containing the drug is placed at the tip, and the tube can be placed over the animal's nose. This method is fairly quick, but is more stressful as it requires some handling prior to sedation. This technique is also useful to extend anesthesia for larger animals during complicated procedures (Lewis, 2004).

Larger mammals require direct injection of drugs into the bloodstream to be properly sedated. Most injectable drugs are considered controlled substances and therefore require special permits, usually issued to licensed veterinarians for acquisition. Injections are delivered with a syringe or more specialized equipment such as tranquilizer darts or a jab-stick. A jab-stick is simply a syringe at the end of a pole; on applying pressure on the end of the pole, the drug is injected into the animal. Mammals in cage traps can be coaxed to move to the end of the trap using a trap divider, which is a removable barrier that is placed through the bars of the trap to confine a captured animal. Confinement of the animal allows less room for it to escape the needle. Animals are usually injected in the flank using this method. Injectable agents work fairly quickly, depending on the dosage, but take longer for the animal to fully recover. Recovery time is an important consideration in the field as drugged animals are vulnerable to potential predators and injury.

9.7 IDENTIFICATION AND MARKING OF ANIMALS

In many cases when studying vertebrates in the field, it is necessary to know the identity of individuals. For example, when estimating population sizes, the number of new captures relative to previously captured individuals is a key to calculating the number of animals in the population. For behavioral studies, it is helpful to be able to discriminate between individuals, both for monitoring while in captivity and for ensuring data collection is not unnecessarily repeated for individuals. As with most field techniques, the researcher must take care that behavior and survival are not negatively affected by the marking process. There are a variety of methods for marking individuals which fall into three groups: natural, noninvasive, and invasive (Silvy et al., 2005); animal welfare should always be taken into consideration when selecting the best technique for marking animals.

In some cases, there is enough natural variation in skin pigments that marking animals is unnecessary. This is preferable to more invasive techniques because animals are not exposed to the stress of handling and marking. These characters include spots and color patterns on amphibians and reptiles and pelage patterns on mammals. These are more easily used for larger diurnal species, where observation of individuals is not hindered by light conditions or visual obstructions.

9.7.1 NONINVASIVE TECHNIQUES

If the researcher only needs to differentiate between marked and unmarked individuals without knowing the identity of each individual, techniques can be used that do not require individual capture. Most of these techniques utilize dyes to mark animals with a remote device to spray the dye on the animal. Birds can be marked with dyes placed on nests or eggs. Dyes generally only last until the animals molt and should not be used for multiyear studies. These techniques can also be useful for species that are difficult to mark, such as shrews. Due to the lack of external pinnae, marking shrews is problematic for population studies. However, marking captured animals using ink, although short-term, is useful for knowing which animals have been previously captured and thus provides some information.

9.7.2 INVASIVE TECHNIQUES

Invasive techniques should be used for cases where more permanent identification is necessary, and consideration should be given to animal welfare. Some studies require researchers to know the identity of individual animals, and marking techniques vary by the amount of stress and discomfort for the animal. As a rule of thumb, the least invasive technique should be used to prevent lasting effects of the marking process, which may alter behavior or survival of the marked animals.

Many different types of invasive marking techniques involve minor mutilation of captured animals. From an animal welfare standpoint, some of these methods can be hard to justify, as they cause pain and stress for the animal, and should only be used when no other alternate method is feasible or available. For birds or mammals, feathers or fur can be clipped or removed. Feather clipping must be used with care as improper clipping can interfere with flight. Fur removal is humane for mammals but relatively temporary, lasting only until the next molt. Hair may be removed using mechanical clippers or chemicals such as depilatory creams. Some chemicals can cause skin irritation for the animal, so researchers should allow recovery time for marked animals. The most common method for marking turtles is notching the shell. However, care must be taken to not weaken the shell. This can be achieved by not notching the shell at the junction of the plastron and the carapace. Snakes are marked by scale clipping subcaudal scales, which leaves permanent scars.

9.7.2.1 Toe-Clipping

A common method of marking terrestrial vertebrates is toe-clipping, in which toes are excised in unique patterns to mark individuals. This technique was more commonplace for small mammals prior to technological advances which rendered toe-clipping obsolete. There is evidence of toe-clipping affecting capture rates for small mammals such as voles (Wood and Slade, 1990), and it is not endorsed by the American Society of Mammalogists (Sikes and Animal Care and Use Committee of the American Society of Mammalogists, 2016). However, it is still used for many amphibians. From a marking perspective, there are several problems with toe-clipping. First, the number of potential combinations is limited by the number of toes for each individual. To reduce the effects of marking, the rule of thumb is to cut no more than one toe per foot. Second, this activity could lead to a bias in the negative effects of marking within populations. To reduce the potential effects on animals, most researchers cut as few toes as possible (one or two per animal), but as more animals are captured, more toes need to be cut (up to one toe per foot). In other words, animals that are captured later lose more toes than those captured initially and suffer greater negative consequences than earlier captures (McCarthy and Parris, 2004). Third, certain species lose toes naturally (or in some cases regrow toes) making reading the specific marks difficult (Silvy et al., 2005). Finally, there is a greater perceived ethical problem with toe-clipping based on an anthropocentric view of the pain endured (May, 2004; Corrêa et al., 2013). For amphibians and some reptiles, toe-clipping has been shown to be less stressful, but it can affect recapture rates for some frog species (Perry et al., 2011). Moreover, studies have demonstrated negative effects of toe-clipping on locomotion for some species (Schmidt and Schwarzkopf, 2010). Although alternative methods can be used to mark other vertebrate groups [such as passive integrated transponders (PITs) and visible implant elastomer (VIE)], toe-clipping still appears to be the most cost-effective technique for marking amphibians (Funk et al., 2005). However, researchers should take care not to remove more toes than necessary (one or two toes ideally; McCarthy and Parris, 2004), and the information benefit can be increased by preserving the toes collected for additional study (i.e., genetic sampling).

9.7.2.2 Visible Implant Elastomer

An alternative to toe-clipping of amphibians is the use of VIE. The elastomer is a two-part compound, mixed in the field and injected with a syringe into the animal to be marked. The liquid elastomer cures into a pliable solid that carries the color with which it was mixed (Northwest Marine Technology, Inc., 2008). A variety of colors and injection locations allow for unique marks (Nauwelaerts et al., 2000; MacNeil et al., 2011). Although more expensive, VIE marking has several advantages over toe-clipping. First, the marks are easily read. There have been some reports of marks moving (Moosman and Moosman, 2006), but in general marked animals are easy to determine. Careful consideration of colors and injection locations can reduce problems with reading marks. Sapsford et al. (2015) suggest using distinctive color combinations (e.g., blue and pink) and avoiding potentially confusing color combinations, such as yellow and green. Moreover, they emphasize the importance of preliminary study to determine which combination of colors and injection points are most effective for reading marks.

Another advantage to VIE marks is that they last a long time. Factors that negatively affect mark permanence are movement of the elastomer under the skin or loss of the mark through the injection point before the mixture cures. A major problem with movement of marks occurs when the marks move to a more darkly pigmented part of the body, where the mark is less distinct (Moosman and Moosman, 2006). Many studies demonstrate the longevity of VIE marks in laboratory and wild populations (Nauwelaerts et al., 2000; Swanson et al., 2013). In addition, most of the problems that occur with VIE marks occur within a short time postmarking. Another benefit of VIE marking, which is especially relevant for the study of amphibians, regards metamorphosis. VIE marks are one of the few marking techniques that can persist through the morphological changes that occur during metamorphosis, so individuals can be tracked through different life stages. In the case of frogs, 67% of wood frogs [*Lithobates (Rana) sylvatica*] maintained their marks through metamorphosis (Grant, 2008) whereas 100% of green and golden bell frogs (*Litoria aurea*) maintained their marks in the laboratory (Bainbridge et al., 2015). However, in the field, the percentage of marked *L. aurea* tadpoles that maintained their marks dropped to 85% (Bainbridge et al., 2015). Also, there is significant variation in how long after metamorphosis marks last, ranging from 2 weeks to 760 days. In general, small marks survive metamorphosis better, but it is advisable to remark juveniles and adults once they have finished the transition.

Finally, there are few ethical considerations with VIE marking. VIE marks are not constrained by the size of the species being tagged and has been successfully used on very small species [e.g., common mistfrogs (*Litoria rheocola*); Sapsford et al., 2014] and individuals of different age groups (juveniles and adults). Compared to the problems associated with toe-clipping, VIE is a more humane technique, when done properly. There is little evidence that indicates a significant negative effect of VIE marking on animal survival (Swanson et al., 2013; Sapsford et al., 2014). Additionally, no effect on locomotion behavior has been demonstrated (Schmidt and Schwarzkopf, 2010; Sapsford et al., 2014). However, many studies on the effect of marking are conducted in laboratory settings, taking behavioral effects out of ecological context (Carlson and Langkilde, 2013).

9.7.2.3 Tags and Bands

A common method for marking birds and mammals is using tags or bands. Tags are usually plastic or metal, with numbers printed or stamped on them for identification of individuals. Placement of tags depends on the species being studied: many bird species are marked with leg bands, whereas most terrestrial mammal species can be marked using ear tags. Numbering on ear tags is determined by the investigator, but leg bands for migratory birds are regulated by the United States Fish and Wildlife Service and the Canadian Wildlife Service. A special banding permit is required for those working with migratory birds. For many behavioral studies, colored bands can be used in different combinations to mark and observe individuals with limited handling. However, studies have demonstrated that the color of the tags themselves affect sexual selection in zebra finches (Burley, 1981), although the consistency of these preferences has recently been challenged (Seguin and Forstmeier, 2012). Regardless, to limit potential behavioral effects of color bands, colors should be chosen that are not visible to species that perceive color.

There are disadvantages to these tags. Most importantly, application requires some handling of the captured animal, which can be stressful. Larger animals may require chemical immobilization prior to marking. Also, animals will generally need to be handled to read the tags. For small mammals, numbers are very small, and mistakes are easily made when tags are hastily read. In addition, tag loss is a problem. This can be ameliorated to some degree by careful placement of the tag within the ring of cartilage at the base of the ear. Silvy et al. (2005) suggest tag placement at the base of the ear and flush with the margin of the ear helps prevent tag loss in white-tailed deer. Also, tagging animals in both ears provides some insurance against tag loss. Tag placement should also take any potential behavioral effects into account. For example, woodland deer mice (*Peromyscus maniculatus gracilis*) have somewhat large ears, so tags are placed at the base of the ear to avoid "sad mouse," in which the weight of the tag pulls the external pinnae down over the mouse's head (Fig. 9.1). This not only makes it easier for the mouse to pull out the tags, but also may have unforeseen effects on movement, avoidance of predators, and overall survival. Despite these minor problems, ear tags are used by many researchers. Ear tags are effective because they are easy to apply, relatively low cost, and provide only momentary discomfort for the animal, if they are applied correctly.

9.7.2.4 Passive Integrated Transponders

One disadvantage of most marking techniques used for mark-recapture studies is the necessity of handling animals every time they are captured to determine their identity. Paired with the need for tag retention, this has led to an increased use of PITs.

FIGURE 9.1

Proper placement of ear tags. If tags are placed along the widest part of the ear, the weight of the tag tends to pull the pinna down, forming "sad mouse" (left). Tags should be positioned at the base of the ear (right), to avoid this from happening.

Ear tags and VIE marks can be lost or become illegible with age; PIT tags resolve both of these issues. PIT tags are small (8—12 mm), consisting of an electronic microchip encased in glass. Transponders are activated by an electromagnetic field generated by a tag-reading device and transmit a unique numeric or alphanumeric code to the reader. Their small size and independence from a constant power source make them very useful, especially for smaller or harder to mark vertebrates. PIT tags are usually injected subcutaneously into each animal. For smaller species, or juvenile life stages, a small incision is necessary to implant the tag. In most cases, surgery is brief (<5 min), and secondary infection is rare.

Tag retention is important for any population study. Retention of PIT tags is high: most tag loss occurs immediately after implantation because the tag slips through the wound made during implantation. Careful treatment of the wound with tissue adhesive can significantly reduce tag loss (Lebl and Ruf, 2010). As PIT tags do not rely on external power and are implanted subdermally, they can last several years. Moreover, the clarity of marks does not change over time, so researchers can be sure that marks are read accurately.

PIT tags have been used in a wide variety of vertebrates, including fish, lizards, snakes, frogs, salamanders, birds, and small mammals (Elbin and Burger, 1994; Gibbons and Andrews, 2004; Fokidis et al., 2006; Bonter and Bridge, 2011). Researchers have addressed potential negative effects of PIT tags on fitness, behavior, and growth, and they found little or no effects (Schroeder et al., 2011; Ousterhout and Semlitsch, 2014; Ratnayake et al., 2014). The only potential problem with PIT tagging concerns frogs. Antwis et al. (2014) found that PIT tagging temporarily disrupted the microbial communities on frog's skin, possibly making them more susceptible to chytridiomycosis, a lethal fungus.

Another advantage of PIT tags is the ability to use them for remote sensing of individuals, which may be difficult to recapture or observe. Because the microchips are only activated in the presence of a chip reader, readers can be equipped with data loggers and placed to record whenever tagged individuals approach within detectable range. Alternatively, detectable ranges can be enhanced with antenna arrays and marked individuals can be located without having to subject them to the stress of capture and handling (Cabarle et al., 2007). This technique is most informative for relatively sedentary species, as it is limited by the detection range of the antennae (usually about 30 cm), and has been used for salamanders in underground burrows (Connette and Semlitsch, 2012; Ryan et al., 2014).

Many researchers have utilized PIT technology in creative ways to address more challenging questions than population monitoring. Not only have PIT tags been used to measure movement in fossorial salamanders (Ousterhout and Semlitsch, 2014), but placement of readers can be used to monitor nest use in birds (Bonter and Bridge, 2011) and predicable microhabitat use (e.g., runways) in mammals (Harper and Batzli, 1996). PIT tags have been used to address questions related to the transmission of disease mediated by behavior (Dizney and Dearing, 2013; Adelman et al., 2014). Other researchers have investigated potential use of habitat corridors by species of interest in conservation efforts (Boarman et al., 1998; Soanes et al., 2015) and the structure of social networks (Nomano et al., 2014). Furthermore, delivery systems are being explored to deliver PIT tags remotely, via implant darts, to be used for larger mammals (Walter et al., 2012).

9.8 CONCLUSION

Despite the difficulties of conducting fieldwork with vertebrates, research projects can be successful and fulfilling with careful planning. The intent of this chapter was not to provide or anticipate every potential problem or outline every technique used in the study of wild vertebrates, but rather to provide a broad overview of some common issues that should be considered prior to commencement of a field study of vertebrates. Much of this chapter heralds from my personal experience over the past 25 years, working mainly not only with small mammals but also with bats, raccoons, opossums, skunks, amphibians, and reptiles. As with any field study, flexibility and ingenuity in problem-solving will lead to many solutions for the unique situations researchers may encounter.

REFERENCES

Adelman, J.S., Moyers, S.C., Hawley, D.M., 2014. Using remote biomonitoring to understand heterogeneity in immune-responses and disease-dynamics in small, free-living animals. Integ. Comp. Biol. 54, 377–386.

Ahumada, J.A., Hurtado, J., Lizcano, D., 2013. Monitoring the status and trends of tropical forest terrestrial vertebrate communities from camera trap data: a tool for conservation. PLoS One 8, e73707.

Animal Care Committee of the Animal Behavior Society, 2012. Guidelines for the treatment of animals in behavioural research and teaching. Anim. Behav. 83, 301–309.

Antwis, R.E., Garcia, G., Fidgett, A.L., Preziosi, R.F., 2014. Tagging frogs with passive integrated transponders causes disruption of the cutaneous bacterial community and proliferation of opportunistic fungi. Appl. Env. Microbiol. 80, 4779–4784.

Awata, S., Tsuruta, T., Yada, T., Iguchi, K., 2013. Stress hormone responses in ayu *Plecoglossus altivelis* in reaction to different catching methods: comparisons between electrofishing and cast netting. Fish. Sci. 79, 157–162.

Bainbridge, L., Stockwell, M., Valdez, J., Klop-Toker, K., Clulow, S., Clulow, J., Mahony, M., 2015. Tagging tadpoles: retention rates and impacts of visible implant elastomer (VIE) tags from the larval to adult amphibian stages. Herpetol. J. 25, 133–140.

Boarman, W.I., Beigel, M.L., Goodlett, G.C., Sazaki, M., 1998. A passive integrated transponder system for tracking animal movements. Wildl. Soc. Bull. 26, 886–891.

Bonter, D.N., Bridge, S.E., 2011. Applications of radio frequency identification (RFID) in ornithological research: a review. J. Field Ornithol. 82, 1–10.

Braun, C.E., 2005. Techniques for Wildlife Investigations and Management, sixth ed. The Wildlife Society, Bethesda, Maryland.

Bub, H., 1995. Bird Trapping and Bird Banding: A Handbook for Trapping Methods All over the World. Cornell University Press, Ithaca, New York.

Burley, N., 1981. Sex ratio manipulation and selection for attractiveness. Science 211, 721–722.

Burnham, K.P., Anderson, D.R., 1984. The need for distance data in transect counts. J. Wildl. Manag. 48, 1248–1254.

Burton, A.C., Neilson, E., Moreira, D., Ladle, A., Steenweg, R., Fisher, J.T., Bayne, E., Boutin, S., 2015. Wildlife camera trapping: a review and recommendations for linking surveys to ecological processes. J. Appl. Ecol. 52, 675–685.

Calisi, R.M., Bentley, G.E., 2009. Lab and field experiments: are they the same animal? Horm. Behav. 56, 1–10.

Carbale, K.C., Henry, F.D., Entzel, J.E., 2007. Experimental analysis of RFID antennas for use in herpetological studies using PIT tagged salamanders (*Ambystoma tigrinum*). Herpetol. Rev. 38, 406–409.

Carlson, B.E., Langkilde, T., 2013. A common marking technique affects tadpole behavior and risk of predation. Ethology 119, 167–177.

Cattet, M.R.L., 2013. Falling through the cracks: shortcomings in the collaboration between biologists and veterinarians and their consequences for wildlife. ILAR J. 54, 33–40.

Chase, S.K., Levine, A., 2016. A framework for evaluating and designing citizen science programs for natural resources monitoring. Conserv. Biol. http://dx.doi.org/10.1111/cobi.12697.

Connette, G.M., Semlitsch, R.D., 2012. Successful use of a passive integrated transponder (PIT) system for below-ground detection of plethodontid salamanders. Wildl. Res. 39, 1–6.

Corn, P.S., Bury, R.B., 1990. Sampling Methods for Terrestrial Amphibians and Reptiles. General Technical Report PNW-GTR-256. U. S. Department of Agriculture, Forest Service.

Corrêa, D.T., Guimarães, T.A., Oliveria, A.L., Martins, M., Sawaya, R.J., 2013. Toe-clipping vital to amphibian research. Nature 493, 305.

Cramer, M.J., Cameron, G.N., 2007. Effects of bot fly, *Cuterebra fontinella*, parasitism on male aggression and female choice in *Peromyscus leucopus*. Anim. Behav. 74, 1419–1427.

Curzer, H.J., Wallace, M.C., Perry, G., Muhlberger, P.J., Perry, D., 2013. The ethics of wildlife research: a nine R theory. ILAR J. 54, 52–57.

Cuthill, I., 1991. Field experiments in animal behaviour: methods and ethics. Anim. Behav. 42, 1007–1014.

Dizney, L., Dearing, M.D., 2013. The role of behavioural heterogeneity on infection patterns: implications for pathogen transmission. Anim. Behav. 86, 911–916.

Dodd Jr., C.K. (Ed.), 2009. Amphibian Ecology and Conservation: A Handbook of Techniques. Oxford University Press, New York.

Elbin, S.B., Burger, J., 1994. Implantable microchips for individual identification in wild and captive populations. Wildl. Soc. Bull. 22, 677–683.

Fair, J., Paul, E., Jones, J. (Eds.), 2010. Guidelines to the Use of Wild Birds in Research. Ornithological Council, Washington, DC.

Farnsworth, E.J., Rosovsky, J., 1993. The ethics of ecological field experimentation. Conserv. Biol. 7, 463–472.

Fellers, G.M., Drost, C.A., Heyer, W.R., 1994. Handling live amphibians. In: Heyer, W.R., Donnelly, M.A., McDiarmid, R.W., Hayek, L.C., Foster, M.S. (Eds.), Measuring and Monitoring Biological Diversity: Standard Methods for Amphibians. Smithsonian Institution Press, Washington, DC, pp. 275–276.

Flagel, D.G., Belovsky, G.E., Beyer Jr., D.E., 2016. Natural and experimental tests of trophic cascades: gray wolves and white-tailed deer in a Great Lakes forest. Oecologia 180, 1183–1194.

Fokidis, H.B., Robertson, C., Risch, T.S., 2006. Keeping tabs: are redundant marking systems needed for rodents? Wildl. Soc. Bull. 34, 764–771.

Fowler, M.E., 1995. Restraint and Handling of Wild and Domestic Animals, second ed. Iowa State University Press, Ames, Iowa.

Funk, W.C., Donnelly, M.A., Lips, K.R., 2005. Alternative views of amphibian toe-clipping. Nature 433, 193.

Gehrt, S.D., Hungerford, L.L., Hatten, S., 2001. Drug effects on recaptures of raccoons. Wildl. Soc. Bull. 29, 833–837.

Gibbons, J.W., Andrews, K.M., 2004. PIT tagging: simple technology at its best. Bioscience 54, 447–454.

Grant, E.H.C., 2008. Visual implant elastomer mark retention through metamorphosis in amphibian larvae. J. Wildl. Manage. 72, 1247–1252.

Guillory, V., 1993. Ghostfishing by blue crab traps. North Am. J. Fish. Manage. 13, 459–466.

Harmon, T.S., 2009. Methods for reducing stressors and maintaining water quality associated with live fish transport in tanks: a review of the basics. Rev. Aquacult. 1, 58–66.

Harper, S.J., Batzli, G.O., 1996. Monitoring use of runways by voles with passive integrated transponders. J. Mammal 77, 364–369.

Hayes, D.B., Ferreri, C.P., Taylor, W.W., 2012. Active fish capture methods. In: Zale, A.V., Parrish, D.L., Sutton, T.M. (Eds.), Fisheries Techniques, third ed. American Fisheries Society, Bethesda, Maryland, pp. 267–304.

Herpetological Animal Care and Use Committee (HACC) of the American Society of Ichthyologists and Herpetologists, 2004. Guidelines for Use of Live Amphibians and Reptiles in Field and Laboratory Research. http://www.asih.org/sites/default/files/documents/resources/guidelinesherpsresearch2004.pdf.

Heyer, W.R., Donnelly, M.A., McDiarmid, R.W., Hayek, L.C., Foster, M.S. (Eds.), 1994. Measuring and Monitoring Biological Diversity: Standard Methods for Amphibians. Smithsonian Institution Press, Washington, DC.

Hubert, W.A., Pope, K.L., Dettmers, J.M., 2012. Passive capture techniques. In: Zale, A.V., Parrish, D.L., Sutton, T.M. (Eds.), Fisheries Techniques, third ed. American Fisheries Society, Bethesda, Maryland, pp. 223–266.

Koprowski, J.L., 2002. Handling tree squirrels with a safe and efficient restraint. Wildl. Soc. Bull. 30, 101–103.

Kunz, T.H., Parsons, S., 2009. Ecological and Behavioral Methods for the Study of Bats, second ed. Johns Hopkins University Press, Baltimore, Maryland.

Kunz, T.H., Tidemann, C.R., Richards, G.C., 1996. Capturing mammals — small volant mammals. In: Wilson, D.E., Cole, F.R., Nichols, J.D., Rudran, R., Foster, M.S. (Eds.), Measuring and Monitoring Biological Diversity: Standard Methods for Mammals. Smithsonian Institution Press, Washington, DC, pp. 122–146.

Lebl, K., Ruf, T., 2010. An easy way to reduce PIT-tag loss in rodents. Ecol. Res. 25, 251–253.

Lewis, J.C.M., 2004. Field use of isoflurane and air anesthetic equipment in wildlife. J. Zoo Wildl. Med. 35, 303–311.

Lobo, N., Green, D.J., Millar, J.S., 2013. Effects of seed quality and abundance on the foraging behavior of deer mice. J. Mammal 94, 1449–1459.

MacNeil, J.E., Dharmarajan, G., Williams, R.N., 2011. SalaMarker: a code generator and standardized marking system for use with visible implant elastomers. Herpetol. Conserv. Biol. 6, 260–265.

May, R.M., 2004. Ethics and amphibians. Nature 431, 403.

McCarthy, M.A., Parris, K.M., 2004. Clarifying the effect of toe clipping on frogs with Bayesian statistics. J. Appl. Ecol. 41, 780–786.

Meador, M.R., 2012. Effectiveness of seining after electrofishing to characterize stream fish communities. North Am. J. Fish. Manage. 32, 177–185.

Meador, M.R., Cuffney, T.F., Gurtz, M.E., 1993. Methods for sampling fish communities as part of the National Water-Quality Assessment Program. U.S. Geological Survey, Open-File Report 93–104. Raleigh, North Carolina.

Meek, P., Fleming, P., Ballard, G., Banks, P., Claridge, A., Sanderson, J., Swann, D., 2014. Camera Trapping: Wildlife Management and Research. CSIRO Publishing, Collingwood.

Minteer, B.A., Collins, J.P., 2005. Why we need an "ecological ethics". Front. Ecol. 3, 332–337.

Moosman, D.L., Moosman Jr., P.R., 2006. Subcutaneous movements of visible implant elastomers in wood frogs (*Rana sylvatica*). Herpetol. Rev. 37, 300–301.

Nations, J.A., Olson, L.E., 2015. Climbing behavior of northern red-backed voles (*Myodes rutilus*) and scansoriality in *Myodes* (Rodentia, Cricetidae). J. Mammal 96, 957–963.

Nauwelaerts, S., Coeck, J., Aerts, P., 2000. Visible implant elastomers as a method for marking adult anurans. Herpetol. Rev. 31, 154–155.

Nomano, F.Y., Browning, L.E., Nakagawa, S., Griffith, S.C., Russell, A.F., 2014. Validation of an automated data collection method for quantifying social networks in collective behaviours. Behav. Ecol. Sociobiol. 68, 1379–1391.

Northwest Marine Technology, Inc., 2008. Visible Implant Elastomer Tag Project Manual: Guidelines on planning and conducting projects using VIE and associated equipment. www.nmt.us/support/appnotes/ape06.pdf.

O'Connell, A.F., Nichols, J.D., Karanth, K.U., 2011. Camera Traps in Animal Ecology: Methods and Analyses. Springer, New York City, New York.

Ousterhout, B.H., Semlitsch, R.D., 2014. Measuring terrestrial movement behavior using passive integrated transponder (PIT) tags: effects of tag size on detection, movement, survival, and growth. Behav. Ecol. Sociobiol. 68, 343–350.

Parker, W.T., Muller, L.I., Gerhardt, R.R., O'Rourke, D.P., Ramsay, E.C., 2008. Field use of isoflurane for safe squirrel and woodrat anesthesia. J. Wildl. Manage. 72, 1262–1266.

Paul, E., Sikes, R.S., 2013. Wildlife researchers running the permit maze. ILAR J. 54, 14–23.

Perry, G., Wallace, M.C., Perry, D., Curzer, H., Muhlberger, P., 2011. Toe clipping of amphibians and reptiles: science, ethics, and the law. J. Herpetol. 45, 547–555.

Porreca, A.P., Pederson, C.L., Laursen, J.R., Columbo, R.E., 2013. A comparison of electrofishing methods and fyke netting to produce reliable abundance and size metrics. J. Freshwater Ecol. 28, 585–590.

Ratnayake, C.P., Morosinotto, C., Ruuskanen, S., Villers, A., Thomson, R.L., 2014. Passive integrated transponders (PIT) on a small migratory passerine bird: absence of deleterious short and long-term effects. Ornis Fennica 91, 244–255.

Rudran, R., Kunz, T.H., 1996. Ethics in research. In: Wilson, D.E., Cole, F.R., Nichols, J.D., Rudran, R., Foster, M.S. (Eds.), Measuring and Monitoring Biological Diversity: Standard Methods for Mammals. Smithsonian Institution Press, Washington, DC, pp. 251–254.

Ryan, K.J., Zydlewski, J.D., Calhoun, A.J.K., 2014. Using passive integrated transponder (PIT) systems for terrestrial detection of blue-spotted salamanders (*Ambystoma laterale*) *in situ*. Herpetol. Conserv. Biol. 9, 97–105.

Sapsford, S.J., Alford, R.A., Schwarzkopf, L., 2015. Visible implant elastomer as a viable marking technique for common mistfrogs (*Litoria rhecola*). Herpetologica 71, 96–101.

Sapsford, S.J., Roznik, E.A., Alford, R.A., Schwarzkopf, L., 2014. Visible implant elastomer marking does not affect short-term movements or survival rates of the treefrog *Litoria rheocola*. Herpetologica 70, 23—33.

Schemnitz, S.D., 2005. Capturing and handling wild animals. In: Braun, C.E. (Ed.), Techniques for Wildlife Investigations and Management, sixth ed. The Wildlife Society, Bethesda, Maryland, pp. 239—285.

Schmidt, K., Schwarzkopf, L., 2010. Visible implant elastomer tagging and toe-clipping: effects of marking on locomotor performance of frogs and skinks. Herpetol. J. 20, 99—105.

Schroeder, J., Cleasby, I.R., Nakagawa, S., Ockendon, N., Burke, T., 2011. No evidence for adverse effects on fitness of fitting passive integrated transponders (PIT) in wild house sparrows *Passer domesticus*. J. Avian Biol. 42, 271—275.

Seguin, A., Forstmeier, W., 2012. No band color effects on male courtship rate or body mass in zebra finch: four experiments and a meta-analysis. PLoS One 7, e37785.

Sikes, R.S., Paul, E., 2013. Fundamental differences between wildlife and biomedical research. LAR J. 54, 5—13.

Sikes, R.S., Paul, E., Beaupre, S.J., 2012. Standards for wildlife research: taxon-specific guidelines vs PHS policy. Bioscience 62, 830—834.

Sikes, R.S., 2016. Animal Care and Use Committee of the American Society of Mammalogists, 2016. 2016 Guidelines of the American Society of Mammalogists for the use of wild mammals in research and education. J. Mammal 97, 663—688.

Silvy, N.J., Lopez, R.R., Peterson, M.J., 2005. Wildlife marking techniques. In: Braun, C.E. (Ed.), Techniques for Wildlife Investigations and Management, sixth ed. The Wildlife Society, Bethesda, Maryland, pp. 339—375.

Snyder, D.E., 2003. Invited overview: conclusions from a review of electrofishing and its harmful effects on fish. Rev. Fish Biol. Fish. 13, 445—453.

Soames, K., Vesk, P.A., van der Ree, R., 2015. Monitoring the use of road-crossing structures by arboreal marsupials: insights gained from motion-triggered cameras and passive integrated transponder (PIT) tags. Wildl. Res. 42, 241—256.

Swanson, J.E., Bailey, L.L., Muths, E., Funk, W.C., 2013. Factors influencing survival and mark retention in postmetamorphic boreal chorus frogs. Copeia 2013, 670—675.

Use of Fishes in Research Committee (Joint Committee of the American Fisheries Society, the American Institute of Fishery Research Biologists, and the American Society of Ichthyologists and Herpetologists), 2014. Guidelines for the Use of Fishes in Research. American Fisheries Society, Bethesda, Maryland.

Walter, W.D., Anderson, C.W., VerCauteren, K.C., 2012. Evaluation of remote delivery of passive integrated transponder (PIT) technology to mark large mammals. PLos One 7, e44838.

Wilson, D.E., Cole, F.R., Nichols, J.D., Rudran, R., Foster, M.S. (Eds.), 1996. Measuring and Monitoring Biological Diversity: Standard Methods for Mammals. Smithsonian Institution Press, Washington, DC.

Wood, M.D., Slade, N.A., 1990. Comparison of ear-tagging and toe-clipping in prairie voles, *Microtus ochrogaster*. J. Mammal 71, 252—255.

Zale, A.V., Parrish, D.L., Sutton, T.M. (Eds.), 2012. Fisheries Techniques, third ed. American Fisheries Society, Bethesda, Maryland.

Zimmerman, B.L., 1994. Audio strip transects. In: Heyer, W.R., Donnelly, M.A., McDiarmid, R.W., Hayek, L.C., Foster, M.S. (Eds.), Measuring and Monitoring Biological Diversity: Standard Methods for Amphibians. Smithsonian Institution Press, Washington, DC, pp. 92—97.

Personnel Safety in the Care and Use of Laboratory Animals

10

M.C. Dyson

The University of Michigan Medical School, Ann Arbor, MI, United States

CHAPTER OUTLINE

10.1 INTRODUCTION

While any workplace harbors potential risks, the animal research enterprise has unique risks that need to be considered and managed for the safety of personnel. Workplace risks associated with animal research include direct injuries such as bites, scratches, and kicks; exposure to animal diseases; the development of allergies and asthma in response to animal allergens; and exposure to biological, chemical, or radiological substances administered to animals as part of research studies (toxins, infectious microorganisms, recombinant DNA, chemicals, radionuclides). Institutions, whether they are academic centers, private companies, or government facilities, are responsible for providing safe working environments for personnel. There are a number of regulations and resources for institutions to utilize to create effective occupational health and safety programs to protect personnel who work with animals in research. However, it should be noted that the best programs involve not only institutional commitment but also the commitment of the personnel to understand their risks, follow recommended safety practices, and report hazardous conditions and behaviors when necessary.

This chapter will review regulations and other resources that provide guidance on the management of risks associated with animal research and occupational health and safety programs for personnel. It will discuss the expectations and responsibilities of the institution, and the organizations and individuals involved in overseeing and participating in animal research and personnel safety. Types of hazards and risks associated with animal research will be described, and recommendations for best practices for management of hazardous activities and prevention of injury or illness will be discussed.

10.2 OCCUPATIONAL HEALTH AND SAFETY PROGRAMS
10.2.1 REGULATIONS AND GUIDELINES

The Occupational Safety and Health Act of 1970 (Public Law 91-596) requires that employers provide a workplace free of recognized hazards. The act empowers the OSHA to create and enforce standards for workplace safety (29 CRF 1910) and requires that personnel comply with these health and safety standards. Safety in the workplace covers a broad range of potential hazards. Specific guidance on safety when working with animals in research is provided in a number of publications provided by the CDC-NIH and the NRC. These include *Occupational Health and Safety in the Care and Use of Research Animals* (NRC, 1997), *Biosafety in Microbiological and Biomedical Laboratories fifth edition* (CDC, 2009), and *Occupational Health and Safety in the Care and Use of Nonhuman Primates* (NRC, 2003). Additionally, *The Guide for the Care and Use of Laboratory Animals* (NRC, 2011) directs institutions that use animals in research to provide occupational health programs for personnel who work with animals or who may be exposed to animal products such as tissues, hair, dander, or waste.

10.2.2 PRINCIPLES OF OCCUPATIONAL HEALTH AND SAFETY PROGRAMS AND RISK ASSESSMENT

Effective occupational health and safety programs must incorporate several basic concepts: knowing hazards present in the workplace; avoiding and controlling exposures to those hazards; compliance and consistency with safety policies and guidelines; thorough record keeping and monitoring of hazards, work practices, and personnel injury; and commitment and coordination by the institution and the organizations and individuals responsible for the program (NRC, 1997). Obviously, identification of workplace hazards will dictate the types of risks and preventative programs that need to be in place in any program. Animal research programs carry the same types of risks that are present in most laboratory situations with the added and often complicated risks of working with live animals, which can cause harm and injury directly; or through the experimental agents that they may carry; or by exposure to infectious, chemical, or radiological materials. Once identified, the degree of risk of the hazard should be evaluated systematically via a **risk assessment**. The process of risk assessment (NRC, 1983) consists of three steps following hazard identification: dose—response assessment (how much causes illness or injury), exposure assessment (the level of exposure in the workplace), and lastly, risk characterization (the magnitude of risk to human health; Table 10.1). A hazardous chemical may not represent risk to personnel if it is used in very small amounts or the animals to which it was administered do not shed hazardous amounts of the chemical, while the same chemical may require additional precautions when used in larger quantities or if not disposed of properly. The same consideration may be applied to potential physical injuries such as bite or scratch from an animal. A bite from a mouse may carry less risk of physical injury to staff than bites from a monkey, and bites from healthy animals may carry less risk than those from animals infected with microbial pathogens as part of a research study. Programs for preventing animal bites would need to take into consideration the species of animals and the types of experiments that the animals are subject to when assessing risks of working with animals. Additionally, as research programs evolve and develop so too may the potential hazards that they involve. Therefore continuous assessment and updating of safety programs are important.

Once hazards are identified and their risks assessed, programs must be created to prevent exposure and subsequent injury and illness. Ideally, prevention of exposure to hazards is the most effective method in preventing injury and illness from occurring. Policies, guidelines, and standards of practice for working with potential hazards must be created. Exposure to hazards should be limited by appropriate facility design, use of safety equipment, and protective clothing. Personnel must be provided information and training on workplace risks, standards of practice designed to prevent exposure to hazards, how to report injuries or exposures, and how to obtain medical care if needed. This is often achieved via a variety of mechanisms such as a formal training course, on-the-job training and signage, and periodic communications.

Table 10.1 Assessment of Risk Associated With Animal-Related Research

Criterion	Possible Classifications	Information Resources
Exposure intensity	High Medium Low Absent	Job profile, environmental health and safety assessment, employee history
Exposure frequency	8 h/week or more Less than 8 h/week No direct contact Never	Job profile, environmental health and safety assessment, employee history
Hazards posed by animals	Severe illness Moderate illness Mild illness Illness unlikely	Institutional veterinarian
Hazards posed by materials used in or with animals	Severe illness Moderate illness Mild illness Illness unlikely	Material safety data sheets; CDC-NIH agent summary statements; radiation, chemical, and biological safety committees, environmental health and safety staff
Susceptibility of employee	Direct threat[a] Permanent increase Temporary increase	Medical evaluation, review of personal medical records
Expected incidence or prevalence	High Medium Low None	Published reports, industry experience
History of occupational illness or injury in the position or workplace	Severe Moderate Mild None	Worker compensation reports, OSHA 200 log
Regulatory requirements	Required for any contact Professional judgment permitted	Environmental health and safety office, consultants, risk managers

[a] *Reasonable probability of substantial harm. Americans with Disabilities Act of 1990 (PL 101-336).*
Reproduced from NRC, 1997. Occupational Health and Safety in the Care and Use of Research Animals. National Academy Press, Washington, DC.

10.2.3 INSTITUTIONAL RESPONSIBILITIES

Institutions that utilize animals in research are required to provide safe workplaces for personnel. As there are many different types of organizations that may utilize animals (e.g., academic centers, private companies, or government facilities), there may be a variety of organizational structures in place to oversee workplace safety programs in these different settings. While a number of federal and field-specific regulations exist, each institution must develop its own internal programs to comply with regulatory expectations, to identify and reduce workplace risks, and to educate and protect personnel from illness and injury (NRC, 1997). For a program of occupational health and safety to function well, the institutional leadership must have a clear understanding of workplace safety issues, provide resources to support programs, and support institutional policies designed to ensure workplace safety. The leadership must also encourage a climate of compliance and collaboration for all organizations and individuals involved in the program.

In most institutions a number of individuals, offices, and departments have important roles in animal research and the occupational health and safety program. Therefore collaboration throughout these divisions is integral for a successful program. Research and/or teaching departments utilize animals in their laboratories and programs. They employ a wide range of students, volunteers, staff, managers, and faculty who often have a great deal of direct contact with animals and their tissues and wastes. Employees of animal care programs also have a great deal of direct and indirect contact with animals and include veterinarians and technicians, husbandry personnel, and individuals who clean facilities, equipment, and caging. Facilities and plant staff who service the building and rooms have indirect contact with animals via wastes or ventilation systems. Institutional regulatory entities such as the Institutional Animal Care and Use Committee and Institutional Biosafety Committee oversee research programs with animals and biological hazards and have important roles in the identification, training, and oversight of personnel using animals in research that are integral to occupational health and safety programs. These departments and divisions must be able to identify potential workplace hazards, educate personnel, and enforce standards for safe working practices. These expectations should apply to anyone working in animal research, including volunteers or students, not just full-time employees. Environmental health and safety departments provide risk assessments for workplace hazards, work in conjunction with individuals in specific areas to help identify risks, create standardized work practices, and provide training and education to personnel. Occupational health services provide health screening before employment or new assignments, periodically monitor personnel for health or hazard exposures, or help to identify and possibly treat illness or injury. For all of these groups, administrative support, including documentation of work expectations and training programs, is essential for program compliance and regulatory oversight.

10.2.4 INDIVIDUAL RESPONSIBILITIES

While the institutions, regulatory bodies, occupational health and safety program personnel, departments, and managers have a significant responsibility in ensuring a safe workplace for personnel, no program can be successful without the compliance and support of the personnel themselves. It is vitally important that personnel understand the risks and hazards that they may be exposed to in the workplace and comply with rules and guidelines for work practices that ensure the safe use of those hazards. These practices are important not only for the safety of the individual working directly with animals but also for the safety of coworkers, family, friends, and the community. For example, animal care and veterinary staff can be inadvertently exposed to hazardous materials used in animals when changing cages or providing medical care. Facility repair and custodial staff can be exposed to allergens or other hazardous materials and agents by working near animal areas, ventilation systems, or improperly labeled waste. Improper work practices such as handling hazardous materials in street clothes can result in contamination of those clothes and subsequent contamination of personal vehicles, homes, or public spaces. Hazardous materials or agents can pollute the environment and water supply if they are improperly disposed into the regular garbage supply or water system. Even materials that are commonly used and considered safe such as pharmaceutical drugs that can and pollute the environment with repeated improper disposal. In addition to understanding hazards and apply appropriate safe work practices, staff should have mechanisms reporting safety concerns to responsible personnel if new hazards are recognized or utilized or if personnel see situations in which the practices of others may create hazardous situations.

10.3 HAZARDS ASSOCIATED WITH ANIMAL RESEARCH

Workplace risks associated with animal research include the same types of risks seen in any work environment, such as exposure to cleaning supplies or laboratory chemicals, physical injuries from machines or other equipment, and ergonomic challenges. Adding animals to the working environment creates a number of unique concerns for personnel. Animals themselves can induce physical injury to personnel by biting and scratching and even by producing excessively loud noises that can damage hearing. Personnel may develop allergies to animals and can be exposed to animal-borne diseases. Finally, research protocols often involve exposing animals to hazardous agents or materials that can potentially cause injury or illness to personnel and the environment (e.g., toxins, infectious agents, chemicals, and radionucleotides) (NRC, 1997). It is important that institutions be aware of these risks and provide appropriate assessment, protection, and education that will allow personnel to work with and care for animals safely.

The majority of this chapter focuses on the most common vertebrate species and groups of animals utilized in research, testing, and teaching. These include rodents, rabbits; domestic animals such as cats and dogs; farm animals such as sheep, goats, and pigs; aquatic animals such as frogs and fish; and nonhuman primates. While some of the more unique animals used in research are not specifically discussed, the principles of oversight of occupational health and safety programs concerning their use will still be applicable.

10.3.1 PHYSICAL HAZARDS

10.3.1.1 Direct Injury From Animals

Bites, scratches, kicks, falls, and ergonomic injuries can all be sustained from directly handling animals. Personnel can also experience crushing injuries from being stepped on or fallen on by large animals such as cows or horses. In the general population, dogs are responsible for the majority of reported animal bites (80%), followed by cats (20%) (Patronek and Slavinksi, 2009). In research settings, rodent bites are frequent, yet underreported due to their often superficial nature. However, bites or other injuries sustained from animals should be treated, reported, and documented no matter which species was responsible for the injury. Personnel can sustain physical injury to tissues and potentially recieve percutaneous exposure to infectious agents (bacteria, virus, fungus) carried naturally by animals or from experimental materials administered to animals as part of research protocols.

The degree of risks of physical injuries from animals can vary by species. Although injury induced by small animals such as mice and rats may cause less conspicuous physical injury, they can still carry infection (e.g., rat bite fever, tetanus) and wounds that should be cleaned and treated. Bites from larger animals may be more physically damaging and may carry more risk of complications or spread of infection. The institutional occupational health group should be aware of the risks associated with the types of animals at the institution and have standard first-aid procedures for bites and other injuries, occupational health-care services for personnel, and reporting and documentation mechanisms to track injuries related to animals.

Ideally personnel should be trained in the appropriate handling and restraint of animals to prevent injury to themselves or the animals. Understanding species-specific behaviors and social cues can help personnel predict reactions of animals and modify approach or handling and restraint techniques when necessary (Lindahl et al., 2013). Personnel training programs should provide information about behaviors expected in species present in the workplace. Most animals are much more amenable to handling when they have had positive socialization with humans in their new environment. Stabilization periods and acclimatization programs that allow animals to adjust to their housing settings and the staff who will care for them are very helpful in reducing stress for the animals and human handlers. A variety of species with a diverse array of behavioral needs may be used in the research setting. Prey species such as rodents can be easily physically manipulated and therefore may represent less potential physical harm to personnel than larger animals. However, attention to appropriate handling is important to prevent stress or injury to the animals themselves. Cats, dogs, and pigs have more complex social needs and also have a greater potential to cause physical injury to personnel. Nonhuman primates are highly intelligent social animals that are often very strong, mischievous, and potentially aggressive. These characteristics make them a danger to themselves and to the individuals who work with them. They can also carry **zoonotic diseases** (diseases resulting from animal pathogens) that are potentially fatal to humans. Accordingly, the use of laboratory primates necessitates extensive personnel training to manage the potential to cause illness and injury to staff and to provide a safe and healthy environment for the animals.

10.3.1.2 Physical and Ergonomic Injury

Physical injury in the workplace can occur in a variety of ways including falls, direct injury from machinery, noises produced by machinery and by animals, high pressure vessels containing gases or steam, and acute or chronic ergonomic injuries. Physical injury can also be induced by technologies specific to the research arena such as lasers, electrocautery, and sources of radiation (NRC, 1997).

Facility and equipment maintenance and housekeeping processes are necessary and important to keeping facilities in good repair and safe for staff and animals. However, the routine processes of movement or breakdown of equipment, the presence of water or cleaning supplies, and changes in planned or expected routes or routines can create potential hazards for trips, falls, and spills that can injure staff or animals. Appropriate notification of changes in facilities, equipment, and temporary changes such as wet floors or inoperable equipment is a very important communication tool that can prevent accidental use of temporarily unsafe areas or equipment in the facility. This can occur via electronic messages, standard schedules, or even hazard signs temporarily placed in relevant areas to notify staff that changes in normal use of the facility must be enacted to prevent injury or allow for equipment or facility repair.

In modern animal facilities, many types of machinery may be encountered that are an important part of maintaining the facility, caring for animals, or running experiments. Cage or tunnel washing machines, autoclaves, high pressure sprayers, high pressure vessels for steam or gases, caging carts and racks, freezers, and incubators area are all common in research and animal facilities. When used incorrectly, or when not functioning correctly, equipment can cause electrical injuries, burns, eye or hearing injury, crushing injuries to limbs, and can even trap staff within the devices themselves. Keeping equipment in good repair is important not only to the function of the equipment but also to prevent injury to the users. Pieces of equipment that have known safety hazards should have fail-safe mechanisms built into their processes to allow for users to immediately stop the function of the equipment in emergency situations. It is important to incorporate testing of safeguards on a regular basis to ensure that they are working correctly. For example, large cage-washing equipment has mechanisms to both stop the cage wash cycle from inside and to allow the door to be opened from the inside if someone is trapped in the washer when it is turned on. These emergency systems can age and fail independent of the washer function, ensuring that they are in good repair is essential. Training personnel on proper use of equipment is a fundamental part of preventing injury related to that equipment. An equally important part of that training is to ensure that personnel understand how to detect when equipment is not working properly, how to notify the appropriate maintenance and facility staff, and how to communicate machinery concerns or malfunctions to coworkers to prevent use until repairs have been completed. Directions for standard use procedures and processes for notification of malfunctioning equipment can be standardized as part of training and kept near machinery for reference and used as needed. Log books documenting equipment testing and repair history can also be valuable for monitoring equipment function.

Personnel may be subject to ergonomic injury from improper lifting, or pushing of heavy loads or animals themselves, and from chronic or repetitive behaviors or motions (Bernard, 1997). As with any hazard, training on proper use of equipment, proper lifting, pushing, and pulling techniques is important to preventing ergonomic injury. Facilities can reduce these types of injuries for their staff by investing in equipment designed to reduce injury such as hydraulic lifts and motorized carts and racks and shelves that reduce the need for stretching or bending (Moore et al., 2011). Also, assessing and updating work tasks that involve repetitive motions that can induce chronic injury, such as cage changing, can help prevent or reduce injuries to personnel.

10.3.1.3 Electrical, Fire and Explosive Hazards

Fires and explosions can result from certain types of chemicals, gases, and other flammable products kept in the facility, from pressurized containers, or in the presence of high concentrations of oxygen, which can perpetuate flammable chemical reactions. The use of 100% oxygen or other gases in pressurized canisters is common in anesthetic procedures. The canisters can cause explosive or physical injury if dropped or punctured. Also the use of tools that can create a spark or intense high temperature, such as electrocautery during surgical procedures, can create a fire hazard in the presence of the oxygen being supplied to the animal under anesthesia. This can result in injury to personnel and animals (Saaiq et al., 2012). The use of appropriate storage and handling of chemicals, gases, and pressurized vessels is important for the prevention of injury. Electrical hazards are also common potential causes of fire or explosions that can be prevented through appropriate use and maintenance of equipment, wires and outlets, and engineering controls such as ground fault interrupters. Safe operational procedures should be employed to prevent electrical or fire hazards.

10.3.1.4 Noise and Hearing Injury

Personnel can sustain hearing damage from loud noises produced by equipment used in the workplace and by the animals themselves, especially pigs and dogs. Exposure to high levels of noise can cause permanent hearing loss. Loud noise can also create physical and psychological stress, reduce productivity, interfere with communication and concentration, and contribute to workplace accidents and injuries by making it difficult to hear warning signals (Goelzer et al., 2001). Safe exposure limits for noise are generally less than 90 dB on a time-weighted average over an 8-h day. If levels exceed 85 dB, personnel need to be enrolled in a hearing conservation program that includes monitoring, hearing testing (audiometric tests), and access to and training for using hearing protection, such as ear plugs or ear muffs (29 C.F.R., 1910.95). It is important that personnel are comfortable reporting concerns about noise levels at their worksite. Personnel who experience ringing or humming in the ears or temporary hearing loss after occupational noise exposure need to report these signs to their occupational health provider. A good rule of thumb regarding noise levels is the need to shout to be heard by a coworker who is an arm's length away; safe noise levels should allow personnel to speak in a normal voice and still be heard at close range (NRC, 1997).

10.3.1.5 Physical Hazards Unique to Research Environments

There are a variety of hazards unique to research environments, especially those in which animals are used, that should be considered in occupational safety programs. Sharp objects such as needles and scalpels ("sharps") are frequently used in animal research and veterinary medicine and represent the potential for physical injury, as well as exposure to toxins, chemicals, or infectious agents. Whenever possible the use of sharps should be avoided and replaced with alternative items. The use of sharps can also incorporate safety mechanisms such as restraint devices or safety covers for needles that reduce or minimize risk. As with any task, training personnel for safe use and disposal of sharps can prevent injury to people and animals (Weese and Faires, 2009). Special equipment used in research settings such as sources of ultraviolet and ionizing radiation, lasers, and electrocautery can all cause injury if not used properly (NRC, 1997). The most common use of ultraviolet radiation is in germicidal lamps used to sterilize surfaces and in water sterilization equipment. Personnel need to have appropriate protection against skin and eye injury that can occur with exposure to UV radiation. Additionally, UV radiation reacts with the vapors of chlorinated solvents and can produce a potent lung irritant called phosgene. These solvents should not be used in the presence of UV-B or UV-C (NRC, 1997). Sources of ionizing radiation such as X-rays and gamma rays are fairly common in medical and research settings. Likewise, exposure of animals or materials used in research of radioisotopes is a common research technique. Different types of ionizing radiation may cause eye or skin damage or be hazardous if ingested or inhaled. Appropriate training to use equipment that produces ionizing radiation, such as X-ray machines, and proper monitoring of personnel for exposure levels are required. The use of radioisotopes can cause additional hazards when they are administered directly into the animal or when they are present in wastes and bedding from animals. The United States Nuclear Regulatory Commission regulates the use of radioisotopes, and investigators must be registered with their institution and appropriately trained. Lasers (light amplification by the simulated emission of radiation) and standard electrocautery devices are also sometimes used in surgical or medical procedures that might be used on animals. They can cause burns or other skin trauma, eye injuries, and potentially fires. Both laser and standard electrocautery devices can also produce fumes and gases that can be harmful for personnel.

10.3.2 CHEMICAL HAZARDS

Chemical hazards from either cleaning products or laboratory chemicals are well recognized as potential sources of injury and illness in personnel (Tan et al., 1999; Takada et al., 2008; Chapot et al., 2009). Chemicals provided to animals during research protocols can secondarily (also) affect animal bedding, animal tissues, and waste. Chemical waste, animal tissue waste, or bedding contaminated with hazardous chemicals that is not properly handled can affect laboratory, veterinary, and cage wash staff; personnel who provide services to the research program (e.g., custodial and facility workers); and ultimately, the community by pollution (contamination) of water and soil. The Occupational Safety and Environmental Health (OSEH) regulations, including The Communication Standard (29 CFR 1910.1200) and the Occupational Exposure to Hazardous Chemicals in Laboratories (29 CFR 1910.1450), require that personnel have access to information on chemicals used in the work environment and that they understand the appropriate handling and disposal of those chemicals.

This information is provided primarily in **safety data sheets (SDS; formerly known as material safety data sheets)** that describe chemical properties, hazards, handling, disposal, postexposure recommendations. All laboratories and worksites using hazardous chemicals are required to have a written **chemical hygiene plan (CHP)**, which includes provisions capable of protecting personnel from the health hazards associated with chemicals in the workplace. Additional guidance for working safely with chemicals can be found in *Prudent Practices in the Laboratory: Handling and Disposal of Chemicals* (NRC, 1995). Chemicals represent a range of risks from properties such as corrosion, flammability (combustibility), explosiveness, and direct toxicity. Toxic effects may be reversible, but in some cases they are irreversible, causing permanent injury. Chemical toxins can be carcinogenic, having a direct effect on organs such as the skin, lungs, kidney, and liver and the nervous system. They can cause allergic reactions, serve as mutagens, and may affect the fetus through material exposure. The effect of a chemical depends on the amount, duration, and frequency of exposure to the agent and can vary with individual sensitivities.

Anesthetic gases can leak out of anesthetic machines or be exhaled directly from animals. Chronic exposure to certain gases has been shown to cause organ damage and increase rates of miscarriage in personnel (Smith, 2010). While modern anesthetic gases such as isoflurane and sevoflurane have less evidence of risk associated with them than previously used agents, it is prudent to educate staff on their proper use and ensure that equipment present to remove or "scavenge" waste anesthetic gases is in working order and used appropriately.

It should be noted that even pharmaceutical agents normally used in humans can represent risk to personnel when they are used in animal research. Drugs such as tamoxifen or chemotherapeutic agents are commonly given to animals in research. They should only be administered to humans in therapeutic doses under a physician's direction. If personnel are inadvertently exposed to these drugs when working with animals, serious injury and illness can occur. Technologies that allow for nanoparticle material production with chemicals pose additional risks by potentially altering a chemical's interaction with biological and environmental systems. Not all of these risks or changes are well understood at this time, but warrant consideration (Conti et al., 2008; Carline, 2014). Risk assessment and CHPs take these issues into account when determining how personnel should work with chemicals. Personnel need to be familiar with the hazardous chemicals in their work areas and follow protocols designed to prevent injury or accidental release of those chemicals.

10.3.3 ALLERGENS

Allergic reactions to animals are among the most common conditions affecting the health of those who work with animals. The most common symptoms include skin rashes, nasal congestion and sneezing, itchy eyes, and asthma. Between 11% and 44% of the individuals working with laboratory animals report work-related allergic symptoms (Bush and Stave, 2003). It is rare but possible for a life-threatening allergic reaction (anaphylaxis) to occur in response to animal allergens. Individual sensitivity, as well as the amount of exposure to animal allergens, is the major risk factor for the development of allergies (Kruize et al., 1997). Inhalation is the most common route of exposure, followed by skin and eye exposures.

Any animal can be a potential source of allergens; however, some animals are more likely to provoke allergic reactions. Rats and mice are the most commonly used species in animal research settings. They produce proteins in the urine (rats and mice) and saliva (rats) that are the major source of allergic reactions in humans. Studies have shown that reducing exposure to rats and mice through appropriate protective clothing (gowns, gloves, masks, or respirators) and use of facility and equipment designs that reduce the amount of air contaminated with allergens, such as ventilated caging and negative pressure air flow, can greatly reduce allergen exposure and the subsequent development of clinical signs of allergic reactions (Schweitzer et al., 2003). Rabbits, dogs, cats, and horses are also widely recognized as sources of allergens. The proteins responsible for producing allergic reactions to them are found in the skin, fur, urine, and saliva.

The development of **laboratory animal allergy (LAA)** can be career ending for personnel working in the research environment, as well as a source of lasting compromised health. The prevention of LAA is the best means of preventing these complications for staff. Reduction of exposure through appropriate facility engineering and equipment and the use of protective gear are important and effective for reducing exposure to allergens and preventing the development of LAA. Medical surveillance programs are also important. Early identification of symptoms of LAA can lead to interventions that might reduce exposure and thereby avoid the long-term health consequences of LAA (Acton and McCauley, 2007).

10.3.4 BIOLOGICAL HAZARDS

Personnel can contract **laboratory-acquired infections (LAI)** through many tasks performed in the laboratory, including work with animals that carry diseases either naturally or through experimental inoculation (Singh, 2009). Reports of LAI have decreased from original surveys performed in the 1970s, and this is likely due to improved awareness, safer work practices, and safety equipment employed in biological research (Singh, 2009). Modern techniques for research animal production and care generally result in healthy animals with a low incidence of infectious diseases. However, exposure to zoonotic agents from naturally infected animals is still possible and should be considered a risk of working with animals. While mice and rats are often reared in specific pathogen free conditions and monitored for known zoonotic pathogens, larger species such as dogs, cats, farm animals, and nonhuman primates may have less defined and controlled disease status. Some of these "larger" species therefore carry higher risks for being naturally infected with zoonotic pathogens. The use of high-quality well-cared-for animals, provision of quality veterinary care, preventative health, and disease screening programs for animals used in research are important factors for protecting the health of personnel who work with animals. Research on the mechanisms, diagnosis, treatment, and prevention of infectious disease is an area of study that commonly utilizes live animal models and therefore represents a potential occupational risk for personnel exposure to those experimental agents. A survey of zoonotic infections in laboratory animal workers from 1999 to 2003 reported a range of zoonotic diseases spanning fungal, viral, bacterial, and parasitic agents (Table 10.2). A list of common zoonoses associated with research animals is provided in Table 10.3. Individuals with the highest risk of exposure are those directly working with the animal or infectious agent in laboratories or as a part of research protocols. However, risk also exists for animal caretakers, veterinary staff, and facility and janitorial staff who may be exposed through working with animals, their bedding, or wastes or animal carcasses (Wedman, 1997). Certain animal models used in research may also be more susceptible to contracting and spreading infectious agents, even those agents that they are unlikely to contract or carry in normal situations. Animals with immune deficiencies, such as severe combined immunodeficiency and athymic nude mice, are often used for transplant or infection studies. Xenotransplantation (transplanting cells, tissue or organs between different species) involves immune suppression and the exposure of animals to tissues and potential infectious agents from those materials (including blood borne pathogens from human cells or tissues). Genetic manipulations of animals or the agents themselves may modulate the sensitivity of the animal or the virulence of the agent (or both) in ways that could increase the hazardous nature of the infectious agent for personnel. Guidance for work with genetically manipulated animals or infectious agents can be found in the *NIH Guidelines for Research Involving Recombinant DNA Molecules* (NIH, 2013). It is important that the institutional occupational health and safety program has information regarding the species of animals and experimental protocols involving infectious agents and genetic manipulations to develop a comprehensive program to reduce risk, design and oversee safe work practices, and monitor for occupationally acquired infections.

The United States CDC–NIH publication, *Biosafety in Microbiological and Biomedical Laboratories* (BMBL), fifth edition (CDC, 2009), is a guideline for biosafety practices and policies for working with infectious agents. The BMBL describes risk criteria for establishing levels of containment, also known as biosafety

Table 10.2 Occupationally Acquired Cases of Zoonotic Disease and Their Source of Exposure From 1999 Through 2003 as Reported by Laboratory Workers

Disease	Source Species	Exposure Type	No. of Cases	No. of Confirmed	No. of Reported
Ringworm[a,b,c]	Dog, cat, rabbit, ox	Skin contact	9	4	5
Q fever[d]	Sheep	Inhalation	2	1	1
Giardia spp.	Dog	Unspecified	2	0	0
Pasteurella spp.	Rabbit, bat	Animal bite (n = 1); needlestick (n = 1)	2	0	2
B virus[a,d]	Macaque	Splash to eyes, nose, or mouth (n = 1); unspecified (n = 1)	2	0	2
Cat scratch disease[c]	Cat	Animal bite	2	2	1
Ectoparasites[d]	Mouse, rabbit	Skin contact	2	1	1
Influenza[b]	Ferret, pig	Inhalation	2	0	1
Rhinovirus	Chimpanzee	Inhalation	1	0	1
Mycobacterium spp.[a]	Guinea pig	Unspecified	1	0	1
Bacterial infection—unspecified	Sheep	Splash to eyes, nose, or mouth	1	0	1
Clostridium difficile	Hamster	Animal bite	1	0	1
Simian foamy virus	Baboon	Anima bite or scratch	1	1	1
Total			28	9	18

[a–d] *Multiple cases of disease in the same person are indicated by matching letters.*
Table used with permission from Weigler, B.J., Di Giacomo, R.F., Alexander, S., 2005. A national survey of laboratory animal workers concerning occupational risks for zoonotic diseases. Comp. Med. 55, 183–191.

levels, for infectious agents. Primary risk criteria are infectivity of the agent, the severity of the potential disease caused by the agent, the transmissibility of the agent, and the type of work being conducted with the agent. There are four biosafety levels in which agents are categorized and each of these levels describes the microbiological practices, necessary equipment and facility features required to work safely with the infectious agents (BSL1−4) and animals potentially infected with those agents (ABSL1−4) (Table 10.4). Biosafety level 1 or animal biosafety level 1 (ABSL-1) requires the basic level of protection and is appropriate for most animals and agents that do not cause disease in normal healthy humans. Biosafety level 2 (BSL2 or ABSL2) is for agents of moderate risk that can cause human disease if ingested, absorbed through mucous membranes, or percutaneous exposure such as injection or entry through a wound. Biosafety level 3 (BSL3 or ABSL3) is for agents with a known potential for aerosol transmission to the respiratory tract or agents that can cause serious and potentially lethal infections. Biosafety level 4 (BSL4 or ABSL4) is for exotic agents with a high risk of life-threatening disease by infectious aerosols. Work with agents that have high risks of infection, (A)BSL-3 and (A)BSL-

Table 10.3 Zoonoses Associated With Commonly Used Research Animals

Species Groups	Zoonoses
Rodents (mice, rats, hamsters, guinea pigs)	Chlamydiosis
	Dermatophytosis
	Ectoparasitism
	Leptospirosis
	Lymphocytic choriomeningitis virus
	Camplylobacteriosis
	Hantaviruses
	Rat bite fever
	Salmonellosis
Rabbits	Dermatophytosis
	Ectoparasitism
	Pasteurellosis
	Salmonellosis
Companion animals (cats and dogs)	Amebiasis
	Brucellosis
	Camplylobacteriosis
	Capnocytophagosis
	Cat scratch disease
	Chlamydiosis
	Cryptosporidiosis
	Echinococcosis
	Ectoparasites
	Giardiasis
	Leptospirosis
	Pasteurellosis
	Q fever
	Rabies
	Salmonellosis
	Strongyloides
	Toxoplasmosis
	Tuberculosis
	Dermatophytosis
Farm animals (sheep, goat, pig, cow)	Balantidiasis
	Brucellosis
	Chlamydiosis
	Camplylobacteriosis
	Cryptosporidiosis
	Echinococcosis
	Ectoparasitism
	Erysipelosis

Continued

Table 10.3 Zoonoses Associated With Commonly Used Research Animals—cont'd

Species Groups	Zoonoses
	Dermatophytosis
	Giardiasis
	Leptospirosis
	Pasteurellosis
	Salmonellosis
	Streptococcus
	Orf
	Q fever
	Tuberculosis
Nonhuman primates	Amebiasis
	Macacine herpesvirus 1 (Herpes B) infection
	Balantidiasis
	Camplylobacteriosis
	Cryptosporidiosis
	Ectoparasitism
	Dermatophytosis
	Giardiasis
	Leptospirosis
	Poxvirus
	Salmonellosis
	Shigellosis
	Simian foamy virus infection
	Strongyloidiasis
	Tuberculosis
	Yellow fever
Aquatic species (frogs and fish)	Chlamydiosis
	Cryptosporidiosis
	Mycobacteriosis (fish handlers disease)

Adapted with permission from Hankenson, F.C., Johnston, N.A., Weigler, B.J., Di Giacomo, R.F., 2003. Zoonoses of occupational health importance in contemporary laboratory animal research. Comp. Med. 53 (6), 579–601.

4, is restricted to specialty laboratories that can provide the appropriate equipment and work practices. Agents that have potential for use as agents of bioterrorism are often classified at the (A)BSL-3 and(A)BSL-4 levels. The use of these agents, called Select Agents, is strictly controlled and overseen by the CDC according to Federal Select Agent Regulations (7 C.F.R. Part 331, 9 C.F.R. Part 121, 42 C.F.R. Part 73). Additional recommended resources for biosafety in the laboratory include *Biosafety in the Laboratory: Prudent Practices for Handling and Disposal of Infectious Materials* (NRC, 1989).

It is possible that microbes carried by animals can be resistant to standard antimicrobial (antibiotic) medications. Antimicrobial resistance in the zoonoses

Table 10.4 Summary of Recommended Animal Biosafety Levels (ABSL) for Activities in Which Experimentally or Naturally Infected Vertebrate Animals Are Used

ABSL	Agents	Practices	Primary Barriers and Safety Equipment	Facilities (Secondary Barriers)
1	Not known to consistently cause diseases in healthy adults	Standard animal care and management practices, including appropriate medical surveillance programs	As required for normal care of each species • PPE: laboratory coats and gloves; eye, face protection, as needed	Standard animal facility: • no recirculation of exhaust air • directional air flow recommended • hand washing sink is available
2	Agents associated with human disease Hazard: percutaneous injury, ingestion, mucous membrane exposure	ABSL-1 practices plus: • limited access • biohazard warning signs • "sharps" precautions • biosafety manual • decontamination of all infectious wastes and animal cages prior to washing	ABSL-1 equipment plus primary barriers: • containment equipment appropriate for animal species • PPE: Laboratory coats, gloves, face, eye, and respiratory protection, as needed	ABSL-1 facility plus: • autoclave available • hand washing sink available • mechanical cage washer recommended • negative airflow into animal and procedure rooms recommended
3	Indigenous or exotic agents that may cause serious or potentially lethal disease through the inhalation route of exposure	ABSL-2 practice plus: • controlled access • decontamination of clothing before laundering • cages decontaminated before bedding is removed • disinfectant foot bath as needed	ABSL-2 equipment plus: • containment equipment for housing animals and cage dumping activities • Class I, II, or III biological safety cabinets available for manipulative procedures (inoculation, necropsy) that may create infectious aerosols • PPE: Appropriate respiratory protection	ABSL-2 facility plus: • physical separation from access corridors • self-closing, double-door access • sealed penetrations • autoclave available in facility • entry through ante-room or airlock • negative airflow into animal and procedure rooms • hand washing sink near exit of animal or procedure room

Continued

Table 10.4 Summary of Recommended Animal Biosafety Levels (ABSL) for Activities in Which Experimentally or Naturally Infected Vertebrate Animals Are Used—cont'd

ABSL	Agents	Practices	Primary Barriers and Safety Equipment	Facilities (Secondary Barriers)
4	• Dangerous/exotic agents which pose high risk of aerosol transmitted laboratory infections that are frequently fatal, for which there are no vaccines or treatments • agents with a close or identical antigenic relationship to an agent requiring ABSL-4 until data are available to redesignate the level • related agents with unknown risk of transmission	ABSL-3 practice plus: • entrance through change room where personal clothing is removed and laboratory clothing is put on; shower on exiting • all wastes are decontaminated before removal from the facility	ABSL-3 equipment plus: • maximum containment equipment (i.e. Class III BSC or partial containment equipment in combination with full body, air-supplied positive-pressure suit) used for all procedures and activities	ABSL-3 facility plus: • separate building or isolated zone • dedicated supply and exhaust, vacuum, and decontamination systems • other requirements outlined in the text

PPE, personal protective equipment.
Reproduced from CDC, 2009. Biosafety in Microbiological and Biomedical Laboratories, fifth edition. U.S. Dept. of Health and Human Services, Washington, DC, pp. 103.

or natural flora of companion and laboratory animals is not well defined; however, the risk exists and should be taken seriously. There are reports of antimicrobial resistant bacteria in pets and laboratory animal colonies (Goo et al., 2012; Davis et al., 2014). Prudent use of antibiotics used in research animals is important for the health of animals and humans alike. Recommended practices include the use of appropriate preventative health-care and aseptic techniques where appropriate to prevent infection and judicious use of antibiotics when necessary. As with any zoonotic agent, personnel should use appropriate personal protective equipment (PPE) and animal handling practices to prevent exposure.

10.3.5 FIELD RESEARCH

The diversity of potential locations, species, and experimental protocols involved in field research with animals makes it very difficult to create centralized standards for occupational safety in these areas. The unique nature of this type of research often places the responsibility for development of safety programs on the principal investigators or laboratories themselves. It is valuable for institutions to have guidelines for the types of safety considerations and educational programs that should be made available to students and personnel involved in field research. Like laboratory-based research, physical hazards from the environment and equipment used in studies represent important safety concerns. In field research, dangerous weather and other environmental conditions further complicate potential safety issues. Risks of drowning, hypothermia, sunstroke, wild or venomous animal encounters, and insect bites are scenarios that must be considered, among a variety of others. The likelihood of remote worksites, personnel working in small numbers or alone, and the potential for limited access to first aid or medical facilities should be considered and planned for. As for all personnel who travel outside of the country, safety issues for travelers should be monitored, and personnel should be trained on how to seek help or even evacuation from politically unstable regions. The animals themselves represent similar hazards to those found in laboratory settings such as bites, scratches, crushing injuries, and the spread of infectious zoonotic disease. Wild animals may be more likely to inflict harm on staff than domesticated research animals, especially if physical handling is necessary for data collection. Staff should be well trained and supervised when working with wild, especially potentially physically dangerous or venomous animals.

10.4 BEST PRACTICES FOR PREVENTION

As discussed in previous sections, the identification and assessment of workplace hazards are essential to creating a safe working environment for personnel in animal research settings. Equally important are the provision of education for departments and individuals on safety issues, recommended safe work practices, and methods to report and rectify safety concerns. Personnel and trained occupational safety professionals alike should routinely evaluate worksites for hazard concerns and ensure that appropriate practices are being employed to prevent illness and injury.

10.4.1 SAFETY PRACTICES (PERSONAL PROTECTIVE EQUIPMENT, STANDARD OPERATING PROCEDURES, EQUIPMENT)

Exposure to hazards is controlled by the use of engineering controls, standard work practices, and the use of **PPE** to prevent exposures (NRC, 1997). Engineering controls can be incorporated into the facility itself or safety equipment used in the facility. Common facility safety designs include ventilation controls that prevent the buildup of fumes or control the escape of air that may contain contaminants (negative pressure airflow) and physical safety features such as gates, fences, and nonslip flooring. Common safety equipment includes chemical fume hoods and biosafety cabinets that allow personnel to work with, or handle, chemical or infectious hazards without direct exposure. For any of this equipment to provide protection to personnel, these devices must be used appropriately and be maintained in proper working order. Regular maintenance and function tests should be employed to monitor safety equipment and prevent breakdowns that might expose staff to hazardous materials or situations. Equipment users should be trained to check equipment for proper function before use and given mechanisms to report concerns with the equipment that is not working properly.

Generally recommended work practices used to prevent or control exposure to hazards can be seen in Table 10.5 and include practices to reduce the number of personnel at risk, reduce exposures from direct or indirect contact, percutaneous exposures, ingestion, and inhalation (NRC, 1997). Specific work practices for particular tasks are often defined by written standard operating procedures (SOPs) that provide detailed instructions on expectations for completing tasks or using equipment. Specific areas or activities that may pose risks to staff working with animals include animal transport and restraint, cage cleaning and waste disposal, issues of personal hygiene (i.e., no eating or drinking or applying cosmetics in animal facilities), or facility housekeeping (removing clutter and cleaning and disinfecting regularly) (NRC, 1997). PPE is the last barrier preventing exposure of personnel to workplace hazards. PPE provides a physical barrier between the individual and potential hazards and is intended to cover skin, clothing, eyes, nose, mouth, hair, and in some cases protect the respiratory tract. PPE recommended for each activity corresponds with the type of hazard and the potential areas of exposure for staff. Gloves and gowns are some of the most commonly used PPE and are often recommended for any work with animals. Other PPE commonly employed includes hair bonnets, face masks or face shields, goggles or safety glasses, and shoe covers. Risk assessments and work patterns should be reviewed to ensure that the PPE and materials selected are appropriate for protecting users. In some settings, cloth or paper gowns are sufficient to prevent contamination of clothing with animal dander. In other settings, personnel may need waterproof coverings because the potential hazards may pass easily through cloth gowns. If personnel are working with animals, which may bite or scratch, PPE

Table 10.5 Recommended Practices for Occupational Health and Safety in Research Settings

Categories of Work Practices	Recommendations for Work Practices
Reducing number of personnel at risk of exposure	Restrict access to the work area.
	Provide warnings of hazards and advise about special requirements.
Reducing exposures by direct and indirect contact	Keep hands away from mouth, nose, eyes, and skin.
	Wash hands when contaminated and when work activity is completed.
	Decontaminate work surfaces before and after work and after spills of a hazardous agent.
	Use appropriate methods to decontaminate equipment, surfaces, and wastes.
	Substitute less hazardous materials for hazardous materials whenever possible.
	Wear personal protection equipment (gloves, gowns, and eye protection).
Practices to reduce percutaneous exposures	Eliminate the use of sharp objects whenever possible.
	Use needles with self-storing sheaths or those designed to protect the user.
	Keep sharp objects in view, and limit use to one open needle at a time.
	Use appropriate gloves to prevent cuts and skin exposure.
	Use puncture-resistant containers for the disposal of sharps.
	Handle animals with care and proper restraint to prevent scratches and bites.
Practices to reduce exposure by ingestion	Use automatic pipetting aids; never pipette by mouth.
	Do not smoke, eat, or drink in work areas used for the care and use of research animals.
	Keep hands and contaminated items away from mouth.
	Protect mouth from splash and splatter hazards.
Practices to reduce exposure by inhalation	Use chemical fume hoods, biological safety cabinets, and other containment equipment to control inhalation hazards.
	Handle fluids carefully to avoid spills and splashes and the generation of aerosols.
	Use inline HEPA filters for protection when using a vacuum system.

Employees should understand the hazards associated with the procedures that they are performing, recognize the route through which they can be exposed to those hazards, select work practices that minimize exposures, and through training and experience acquire the discipline and skill necessary to sustain proficiency in the conduct of safe practices.

Reproduced from NRC, 1997. Occupational Health and Safety in the Care and Use of Research Animals. National Academy Press, Washington, DC.

may need to be sturdier to prevent it being ripped or removed altogether during animal restraint. Face shields or safety glasses may not provide sufficient splash protection in certain settings and sealed goggles may be preferred. If personnel need respiratory protection due to chemical, infectious, dust or allergy concerns they should be enrolled in an OSEH compliant respiratory health program (NRC, 1997).

10.4.2 EDUCATION

Educational programs regarding occupational safety concerns and best practices can come in many forms. The best format for providing information varies with the risks present, the complexity and novelty of the information, and the experience and understanding of the personnel. It is valuable to keep detailed records on safety training provided to staff so that areas for improvement or updates can be targeted based on the information provided in previous training sessions. Introductory materials and brief updates and changes can be easily and effectively communicated via email, electronic posts or bulletins, and e-learning modules. Training on appropriate work practices in the laboratory or animal facility should be provided in person by a trainer or experienced staff member so that personnel can observe actual practices and then have their own application of those practices observed until they are competent to perform them alone. Periodic reassessment of skills and techniques at the worksite is also a valuable tool to ensure that practices continue to be employed appropriately. Notices and reminders regarding required safety practices should be posted and regularly updated so that personnel have access to them on a routine basis. Finally, personnel need to be educated on methods of recognizing and reporting hazards and unsafe work practices in their worksites and encouraged to do so. It is important for supervisors and managers to respond to and follow up on feedback and concerns from personnel regarding safety issues so that new or previously unrecognized risks can be assessed, removed, or managed as necessary.

10.4.3 EMERGENCY AND DISASTER PLANNING

While it is ideal that the protective equipment and practices in the workplace prevent any injuries or hazard exposure, it is prudent to plan for unexpected disaster situations that make standard work practices difficult. Similar to a risk assessment, plans for handling potential hazard exposures and the need for altered work practices during a disaster should be considered and the consequences defined. Mechanisms should be in place for personnel to effectively report safety concerns, mistakes, exposures, injuries and illness. For exposure to hazards, personnel should be educated on the proper use of first aid or exposure treatment equipment and how to access urgent or emergency health-care services. Workplaces should contain appropriate first response equipment when necessary and provide training and directions on how to use such equipment, such as eye wash stations, emergency showers, first aid kits, animal bite and scratch kits, and sharps exposure SOPs. Similarly, weather (snow, flood, tornado) and facility (fire, flood, loss of electricity) disasters may shut down normal routes of access to animals or laboratories, release hazards for animals, make deliveries of supplies difficult, and cause additional physical hazards for personnel. Plans for disaster response, altered work practices to contain hazards, and communication with emergency response personnel regarding hazards are important considerations. Training for personnel on emergency or disaster plans is a valuable tool for an effective and safe response.

10.5 CONCLUSIONS: PERSONNEL ENGAGEMENT AND AWARENESS

Occupational safety is a complex and dynamic field, especially with the added challenges of live animals, infectious agents, and other potentially hazardous research tools. The institution is responsible for providing a safe working environment for its personnel. Key elements of an effective occupational health and safety program include administrative procedures, facility design and operation, exposure control, education and training, occupational health services, equipment performance, information management, and program evaluation (NRC, 1997). However, even an ideally designed and administered program is not effective without the engagement of the personnel themselves. Personnel need to be engaged in their work environment and aware of hazards and be knowledgeable of safe work practices. It is important for personnel to know how to report safety concerns with facilities and work practices (their own as well as others). Personnel should also have appropriate mechanisms for reporting changes in their own health status so that work assignments may be altered if their health status puts them in a higher risk category (e.g., pregnancy, immune compromise). Personnel are often the first to see changes or safety concerns arise, and it behooves supervisors, managers, and occupational health professionals to seek the insights of personnel when assessing risks and developing strategies to manage those risks.

REFERENCES

Acton, D., McCauley, L., 2007. Laboratory animal allergy: an occupational hazard. AAOHN J. 55, 241−244.

Bernard, B., 1997. Musculoskeletal Disorders and Workplace Factors. CDC-NIIOSH, Cincinatti.

Bush, R.K., Stave, G.M., 2003. Laboratory animal allergy: an update. ILAR J. 44, 28−51.

Carline, D.J., 2014. Nanotoxicology and nanotechnology: new findings from the NIEHS and Superfund Research Program scientific community. Rev. Env. Health 29, 105−107.

CDC, 2009. Biosafety in Microbiological and Biomedical Laboratories. U.S. Dept. of Health and Human Services, Washington, DC.

Chapot, B., Secretan, B., Robert, A., Hainaut, P., 2009. Exposure to hazardous substances in a standard molecular biology laboratory environment: evaluation of exposures in IARC laboratories. Ann. Occup. Hyg. 53, 485−490.

Conti, J.A., Killpack, K., Gerritzen, G., Huang, L., Mircheva, M., Delmas, M., Harthorn, B.H., Applebaum, R.P., Holden, P.A., 2008. Health and safety practices in the nanomaterials workplace: results from an international survey. Env. Sci. Technol. 42, 3155–3162.

Davis, J.A., Jackson, C.R., Fedorka-Cray, P.J., Barrett, J.B., Brousse, J.H., Gustafson, J., Kucher, M., 2014. Carriage of methicillin-resistant staphylococci by healthy companion animals in the US. Lett. Appl. Microbiol. 59, 1–8.

Goelzer, B., Hansen, C., Sehmdt, G., 2001. WHO|Occupational Exposure to Noise: Evaluation, Prevention and Control, Federal Institute for the Occupational Safety and Health (Germany).

Goo, J.-S., Jang, M.-K., Shim, S.-B., Jee, S.-W., Lee, S.-H., Bae, C.-J., Park, S., Kim, K.-J., Kim, J.-E., Hwang, I.-S., Lee, H.-R., Choi, S.-I., Lee, Y.-J., Lim, C.-J., Hwang, D.-Y., 2012. Monitoring of antibiotic resistance in bacteria isolated from laboratory animals. Lab. Anim. Res. 28, 141–145.

Hankenson, F.C., Johnston, N.A., Weigler, B.J., Di Giacomo, R.F., 2003. Zoonoses of occupational health importance in contemporary laboratory animal research. Comp. Med. 53 (6), 579–601.

Kruize, H., Post, W., Heederik, D., Martens, B., Hollander, A., Van der Beek, E., 1997. Respiratory allergy in laboratory animal workers: a retrospective cohort study using pre-employment screening data. Occup. Env. Med. 54, 830–835.

Lindahl, C., Lundqvist, P., Hagevoort, G.R., Lunner-Kolstrup, C., Douphrate, D.I., Pinzke, S., Grandin, T., 2013. Occupational health and safety aspects of animal handling in dairy production. J. Agromed. 18, 274–283.

Moore, S., Torma-Krajewski, J., Steiner, L., 2011. Practical Demonstrations of Ergonomic Principles. CDC-Centers for Disease Control – NIOSH, Pittsburg.

NIH, 2013. NIH Guidelines for Research Involving Recombinant or Synthetic Nucleic Acid Molecules. (78 FR 66751).

NRC, 1983. Risk Assessment in the Federal Government: Managing the Process. National Academies Press, Washington, DC.

NRC, 1989. Biosafety in the Laboratory: Prudent Practices for Handling and Disposal of Infectious Materials. National Academies Press, Washington, DC.

NRC, 1995. Prudent Practices in the Laboratory: Handling and Disposal of Chemicals. National Academies Press, Washington, DC.

NRC, 1997. Occupational Health and Safety in the Care and Use of Research Animals. National Academy Press, Washington, DC.

NRC, 2003. Occupational Health and Safety in the Care and Use of Nonhuman Primates. National Academies Press, Washington, DC.

NRC, 2011. Guide for the Care and Use of Laboratory Animals, eighth ed. National Academies Press, Washington, DC.

Patronek, G.S., Slavinksi, S.A., 2009. Animal bites. J. Am. Vet. Med. Assoc. 234, 336–345.

Saaiq, M., Zaib, S., Ahmad, S., 2012. Electrocautery burns: experience with three cases and review of literature. Ann. Burns Fire Disasters 25, 203–206.

Schweitzer, I.B., Smith, E., Harrison, D.J., Myers, D.D., Eggleston, P.A., Stockwell, J.D., Paigen, B., Smith, A.L., 2003. Reducing exposure to laboratory animal allergens. Comp. Med. 53, 487–492.

Singh, K., 2009. Laboratory-acquired infections. Clin. Infect. Dis. 49, 142–147.

Smith, F.D., 2010. Management of exposure to waste anesthetic gases. AORN J. 91, 482–494.

Takada, S., Okamoto, S., Yamada, C., Ukai, H., Samoto, H., Ohashi, F., Ikeda, M., 2008. Chemical exposures in research laboratories in a university. Ind. Health 46, 166–173.

Tan, Y.M., Diberardinis, L., Smith, T., 1999. Exposure assessment of laboratory students. Appl. Occup. Env. Hyg. 14, 530–538.

Wedman, A., 1997. History and epidemiology of laboratory-acquired infections (in relation to the cancer research program). J. Am. Biol. Saf. Org. 2, 12–29.

Weigler, B.J., Di Giacomo, R.F., Alexander, S., 2005. A national survey of laboratory animal workers concerning occupational risks for zoonotic diseases. Comp. Med. 55, 183–191.

Weese, J.F., Faires, M., 2009. A survey of needle handling practices and needlestick injuries in veterinary technicians. Can. Vet. J. 50, 1278–1282.

Thesis Development

11

J. Robichaud

University of Notre Dame, Notre Dame, IN, United States

CHAPTER OUTLINE

Once a student's experiments are performed and the data collected, the next step in the scientific process is to prepare the thesis. Although this process can feel like a colossal undertaking, it is a necessary and an important step. A thesis is a set of works that demonstrates that a body of knowledge has been acquired through education and a research program. It is developed to contribute to the general body of work on a given subject. The thesis is written to demonstrate the creativity that went into the research design and the understanding of current research methodologies. It also displays that the student is able to communicate the research findings to the scientific community.

The goal of this chapter is to present a set of general guidelines that the student can use to develop a thesis. All sections of the thesis require the use of a library to access journal articles from others in the field, including to provide historical context for the research, to help develop a hypothesis, to offer procedural alternatives, and to support or refute the data collected. Some of the exact formatting details needed to present the argument or thesis can vary depending on each particular research institution or journal to which the research will be submitted for publication. However, the foundation or structure of a thesis is fairly standard.

11.1 PREPARATION
11.1.1 LITERATURE SEARCH

During **thesis development** of a thesis, a comprehensive search of databases and other library resources is essential to complete the project. The author should refer primary literature, such as journals, dissertations, and conference papers, rather than secondary literature (e.g., encyclopedias, Wikipedia, textbooks, etc.) (Knisely, 2009). Knowing how and where to start this process can cause distress because of the vast number of journal articles that are actually available. Often times, student database searches are too broad and return hundreds to thousands of articles; or equally, so specific, that too few papers are retrieved.

To help the process, a strategy to search for articles should be employed. The following steps will assist students as they begin their venture into building a literature search (University of Notre Dame, 2016):

1. A brief summary of the topic should first be written; one to two sentences to describe the goal of the project.
2. Based on the summary, the key concepts must be identified. It must be determined which concepts or key words and phrases are related to the project. Defining at least two concepts will improve the search results in the next step.
3. For each concept, related terms or synonyms should be established to account for other ways that the concept could be expressed. There may be broader or more concise terms to use in the search, including the use of layperson terms versus scientific terms.
4. There are numerous databases available for researching the thesis topic; some databases are free while others require a subscription. All are continuously updated. Each database has pros and cons. For those performing medical-based research, PubMed (http://www.ncbi.nlm.nih.gov/pubmed) is the most widely used database, which offers various filters to limit the search parameters and does not require a subscription. Other databases, such as Google Scholar, Web of Science, and CAB Abstracts, offer the ability to filter searches to various degrees, may require subscription (university libraries usually hold several subscriptions at once), offer a means to expand the bibliography based on common references, as well as provide other assorted tools depending on the particular database. Although there is a fair

amount of overlap in journal articles between databases, there will also be unique journals in particular databases. Several databases should be accessed to ensure that all resources are discovered (Knisely, 2009).

5. Once a search is completed, it should be examined and refined as necessary. If the search results return too many (>250) or too few (<25) articles, an examination of the search terms used may reveal that a word or phrase may need to be added to narrow the search or removed to broaden it. If the search did not reveal articles that were relevant to the topic, a new search with new concepts (step 2) and key terms (step 3) may need to be completed.

It is imperative to keep the literature organized. After all the work in obtaining articles, losing those articles due to disorganization only makes the process harder. The use of one of the several software programs, such as EndNote or RefWorks, will help to manage the references.

11.1.2 HYPOTHESIS DEVELOPMENT

Before even beginning the research project or the writing process, it is necessary to develop a hypothesis. As a scientist, even a student scientist, asking questions about why and how different events occur in the world is integral to the process of science. Reading primary and secondary literature will assist in answering questions and developing new questions. Once the information has been gathered, then the questions can evolve into a **hypothesis** or a possible explanation for processes that have been observed (Knisely, 2009). It is on this hypothesis that the research proposal will be based, the experimental plan developed, the data analyzed, and the results interpreted.

11.2 THESIS SECTIONS

A thesis is written to include the following sections: introduction, methods and materials, results, and discussion. Each section provides details that support the hypothesis based on the collection of the work.

11.2.1 INTRODUCTION

The **introduction** provides the background about the subject of the thesis. The background, although broad, is important because it establishes the historical framework for understanding the chosen topic more deeply (Evans et al., 2011). By delving into previous research studies, the student is able to provide the audience and the mentor or committee an opportunity to see that the topic was thoroughly investigated. In some circumstances, the introduction is written prior to starting any experiments as part of a research proposal and can influence the development of the project.

The format of the introduction allows the student to tell a compelling story about the project. The information should work from a broad to a narrow focus. Four general questions should be addressed in the introduction: (1) why should the reader care about the research/topic, (2) what is already known, (3) what is not known, and (4) what was accomplished.

The introduction should entice the reader to care about the project. Whether the experiments involve infectious disease, cancer biology, or any other valuable scientific finding, the reader must be provided with an understanding of why the particular topic is important to study. The content can vary but should include some general background. For instance, if the research involves a bacterial or viral infection of animals, there should be basic microbiological information about the organism itself including the type of animals that are affected, seasonality infection rates if they apply, symptomatology, morbidity and mortality rates, and if it is **zoonotic** (able to pass between animals and humans).

Once the basic information on the topic is presented, the next segment should begin to gradually narrow the introduction to explain what is currently known and still unknown about the topic. This can be difficult in some cases, especially if the work is at the forefront of scientific research. However, knowledge from other researchers' experiences in other related fields is used to assist with the explanation as to why it is justified to delve into this topic through this project. Often it is necessary to distinguish between the known and unknown components of a project. For instance, if a particular protein or gene in a novel microorganism has been revealed during the project it is likely that very little is known about the microorganism. However, in this case, it is only the microorganism that is novel, not the protein or gene, thus the information on the known can be focused on the protein that has been found in other potentially related microorganisms where its function has already been elucidated.

The final section of the introduction becomes very specific about the actual research project being presented. The objective and hypothesis are stated here to provide the reader the goals of the project and the student's explanation of the research outcome. Depending on the format of the paper, it is sometimes appropriate to add a brief summary of the results. The author may also provide a basic list of the experimental techniques that were employed to reach the conclusions without the explanation of the theory behind the techniques or the explicit details of them.

11.2.2 MATERIALS AND METHODS

The **materials and methods** section of the thesis contains the basic information of the techniques performed to carry out the experiments. It is essential to provide sufficient details to allow the work to be replicated, but not all of the specifics that a scientific audience would already understand. Although it seems a relatively simple section to write, the materials and methods section can prove difficult for a student for a variety of reasons.

Because there are often numerous techniques used and several stages of the experiment, it can be difficult to present the material in an organized fashion. One way to provide organization to this section is to group the techniques under subheadings. **Subheadings** allow the author an efficient way to place the experiments in an order that makes sense to the reader. Yet, there must also be a logical flow to this section. For instance, reporting all of the gel electrophoresis methods under one subheading may seem logical. However, if the technique is used throughout various stages of the research, it may be necessary to present each method used under different subheadings to maintain the proper sequence of events. The subheadings should be created from the natural sequence of how the research was conducted.

Providing too much information in the materials and methods is a second issue that many thesis writers face. The thesis is written for a scientific audience who should be able to understand scientific terminology and common techniques such as the use of centrifuges, vortexes, and pipettes. It is not necessary to include every single technical detail; just enough information that would allow the reader to repeat the experiment will be sufficient. When kits or newly published techniques are used, a simple citation may be warranted. When cloning a plasmid into a commercially available *Escherichia coli* bacterial strain, for example, the manufacturer's instructions should not be rewritten verbatim into the thesis. Instead, the name of the plasmid and that it was cloned into *E. coli* from a particular kit should be noted, with the company name and city referenced in parentheses. Other considerations for this section include maintaining a consistent verb tense throughout the section, limited or no use of first person, and the removal of extraneous words such as first, next, or then.

11.2.3 RESULTS

The **results** section of a thesis is where a summary of the results are presented in an unbiased approach that illustrates the trends seen in the experiment but without any interpretation (Ambrose et al., 2007). The results section consists of two parts: **figures** and **text**, both being equally important to the thesis. The two subsections should be linked as each part fulfills the requirements of the overall section, but there is often redundancy. However, the text portion provides an objective explanation of the information represented in the figures' portion. All of the collected data must be evaluated and organized so that they are presented in a logical sequence. The mechanics of this work will consist of analyzing the raw data into summarized observations (e.g., statistical analysis, analyze gel images, etc.) and deciding on the best way to represent those observations (type of figure to use) to the reader (Knisely, 2009; Pechenik, 2013).

11.2.3.1 Figure Subsection

Not all data collected are critical to the overall presentation of the research. Trying to present all data will cause confusion to the reader and a loss of focus in the thesis. Choosing which data to present and the graphing technique to use are crucial and should be carefully organized to facilitate the progression of information throughout the paper.

Once graph choices have been established, the paper can be formatted. It is essential to investigate the formatting requirements of each institution or journal as they often have their own particular formatting rules on where to place figures and captions within the results section. In some cases, all of the figures are presented at the end of the results text section, embedded within the text or in an appendix. Regardless of the placement of the figures, the captions should be written with ample explanation that allows for the data being presented to be understood without the accompanying text subsection (Evans et al., 2011).

General rules for creating good figures and properly written captions include the following:

1. **Figure or Table**: The optimal choice for presenting the data to allow for visual representation of the results is essential. The research may have numerical data that require computations and statistics or images that require analysis. For images, the only option is to use the photograph. However, for numerical data, there is a wide array of graphing schemes and table options.
2. **Label number**: All figures and tables must be designated with a number, such as figure 1 or table 2, for reference in other sections of the thesis. This number is normally positioned under the actual figure or table and to the left of the title.
3. **Title**: The title should be concise, yet descriptive. The reader should be able to gather from the title alone what type of experiment was performed and what type of sample was used.
4. **Methods statement**: A brief description of the techniques used to obtain the information on the table or figure should be presented. It should not be verbatim of the materials and methods section but, instead, a brief summary.
5. **Result**: This is a general statement, but not a discussion on the meaning, about the particular figure that is presented. This can include a trend or statistical data about the result.
6. **Formatting**: The formatting will vary with each figure, depending on the type of data being presented. However, some general rules will remain constant.
   a. Additional text, such as a figure legend or axis labels, should be large enough to read and be of a consistent size throughout the document.
   b. All graphs must have appropriate labels and units on the axes.
   c. A legend should be provided to explain different symbols. These can be explained in the methods statement but the efficiency of doing so can vary from figure to figure. If a symbol is used, consistency in all figures is important.
   d. Appropriate statistics should be presented, including standard error bars. The statistical analysis can be written in the results statement of the caption. Another efficient way of presenting the statistics is to add the information in the white space of the graph so that the reader does not need to search other sections of the thesis to determine if the results were significant.

11.2.3.2 Text Subsection

While the figures provide the reader with a visual representation of the analyzed data, the text section summarizes the major observations and trends that were presented in the figures. This section often proves difficult for many writers after having already provided much of the information about the observations and trends in the figure subsection. The largest challenge of the text subsection is to discover new and interesting ways of expressing the findings.

The flow of the text subsection should mirror that of the figure subsection. For the text subsection, the major findings of each figure or table are summarized, in past tense, in individual paragraphs. Usually, each paragraph would begin with a brief description or summary of the experiment without recounting of the materials and methods section (Ambrose et al., 2007). Specifically, the author presents each paragraph as a new set of data, its accompanying statistics, and the parenthetical reference to the figure or table in order of appearance.

The overall wording of the results section should be concise and not contain unnecessary or extraneous wording (Knisely, 2009). Although this may result in a shorter results section, providing additional figures to lengthen the result section will only distract the reader from the focus of the major trends present in the data. Also, there should be no discussions about the implications of the findings in this section. The results section is only the impartial revealing of the data collected in the experiments.

11.2.4 DISCUSSION

The **discussion** section allows the student an opportunity to demonstrate that the experimental results have been accurately analyzed and appropriately interpreted. These conclusions should be drawn in context to the proposed hypothesis and also discussed in relation to the broader issues of the topic (Pechenik, 2013). A well-structured discussion will mirror the introduction by beginning narrowly and ending with the broader impact of the research. A convincing discussion section will contain evidence of what was expected to be found, including the reasoning of this expectation, a comparison of the actual results to the expected results, an explanation of unexpected results, a direction for future work, and a broad concluding statement.

A strong discussion will begin with a general statement that reflects back to the hypothesis found in the introduction. The purpose for restatement is to remind the reader of the original intentions of the project and make a broad statement about the findings (Knisely, 2009). Authors can often find this to be a difficult task as they had already stated the hypothesis in a clear and concise manner in the introduction. However, this stage of the discussion is important because now the interpretation of the results is either to support or to refute the hypothesis.

Obtaining results that exactly support the proposed hypothesis rarely occurs. It is often necessary to explain unexpected results. Even when the research was conducted after proper planning and with all variables considered, unexpected findings can occur. These unpredicted outcomes can lead to important scientific discoveries and thus must be reported and considered. The data collected should be thoroughly analyzed to allow the author to provide a compelling interpretation of these findings, which can then lead to further investigation of the discovery (Pechenik, 2013).

It is important to have explanations made in the context of biology or what is the biological explanation. For example, it should be explained why the author believes that a particular cancer cell line reacted to a drug once it was introduced at various dosage levels and what the mechanism by which it does so. To help develop a biological explanation, delving into the scientific literature may again be required. It may not always be possible to find papers that most closely match the research to help answer those explanations. However, if a previously unknown protein in a bacterium is discovered, then it is possible that researching closely related or even unrelated bacteria with that protein may help to provide insight into the biological explanation in this particular bacterial strain.

Research never ends with the results of one project. There are always unanswered or new questions that develop during the course of a project. The future direction a research project or program takes is usually influenced by the last experiments completed. For all of those unexpected and unanswered questions, the author can propose the next steps to take to understand the biological mechanisms under investigation. The discussion should end with a broad statement or summary paragraph of the overall impact that this new knowledge could have on the world (Knisely, 2009).

11.3 FINALIZING THE THESIS

There are a number of steps to take to finalize a thesis.

11.3.1 PROOFREADING

The word processing program used for the preparation of the thesis will find most of the spelling and grammatical errors. However, it might not able to understand the context or content. The following questions should be answered by the writer: (1) does the paper flow logically to reveal the research, (2) does the material and methods section provide enough information to allow the experiment to be repeated, without providing every miniscule detail, (3) is there limited use of passive voice, first person, past tense, and possessive pronouns, and (4) were quotes avoided. When citing information from another resource, it is wise to paraphrase rather than quote (Pechenik, 2013).

11.3.2 PEER REVIEW

The peer review procedure is an extremely critical aspect of the writing process. Fellow students, technicians from the research laboratory, and the student's mentor can all be called on to review the collection of work. This review includes critique of the writing style, the research experimentation, the analysis of the data, and the overall thought process of the project. Although a difficult process for both the reviewers and the students, a good peer review can elevate the level of scientific writing for the student.

Before the paper is submitted for review, all spelling, grammatical, and sentence structure errors should be corrected so that the reviewer does not lose focus on the content of the thesis. However, some of these errors may be missed as it is difficult for a writer to achieve enough distance from his or her own work to be able to find all of the errors. Another positive outcome of the review process is that the student is provided feedback on the quality of the writing ability as scientific writing is a learned skill that requires experience and time to hone (Pechenik, 2013).

11.3.3 TITLE

Because the title is the first thing from the thesis that readers and other scientists who are performing a literature search in the specific field of science will view, it must be both concise and descriptive. The title should convey the nature of the work that was completed in the research. It is best to wait until the writing has been completed to decide on the title. In determining a title, several questions should be considered: (1) what was studied or what was the independent variable, (2) if an organism was utilized, what is the scientific name of the organism and does it need to be in the title, (3) was there a particular method employed, and (4) what was the goal of the project or what was the response in the dependent variable.

11.3.4 ABSTRACT

The **abstract** is a summary of the entire paper written in no more than 250 words. Each section of the thesis (introduction, materials and methods, results, and discussion) is represented within the abstract. There are no references to figures, literature, or introduction of new data in the abstract.

11.3.5 ACKNOWLEDGMENTS

This section allows recognition of support that was provided to the student by the mentor and any other individuals during the research, writing, and review process. Acknowledgment of a foundation or institute that may have provided funding is also appropriate.

11.3.6 LITERATURE CITED

All literature that was reviewed during this entire process should be cited in this section. All items listed must be referenced within the thesis body. The format to use varies among institutions and journals.

11.4 CONCLUSION

Although the development of a thesis is a daunting task, it is necessary to share the scientific knowledge attained with others in the field. Preparation of the methods and materials section adequately to allow others to duplicate the experiments is crucial. The overall process prepares a student for continued success in the scientific community.

REFERENCES

Ambrose III, H.W., Ambrose, K.P., Emlen, D.J., Bright, K.L., 2007. A Handbook of Biological Investigation, seventh ed. Hunter Textbooks, Inc., Winston-Salem.

Evans, D.G., Gruba, P., Zobel, J., 2011. How to Write a Better Thesis, third ed. Springer, Heidelberg.

Knisely, K., 2009. A Student Handbook for Writing in Biology, third ed. Freeman, New York.

Pechenik, J.A., 2013. A Short Guide to Writing About Biology, eighth ed. Pearson Education, New York.

University of Notre Dame First Year Studies Program, 2016. Starting Your Research: Getting Started. A Beginning Scholar's Guide to Research Success for New Students at the University of Notre Dame. http://libguides.library.nd.edu/first-year-studies/getting-started.

Glossary of Terms and Acronyms

Agencies Acronyms

AAALAC   Association for Assessment and Accreditation of Laboratory Animal Care, International
AALAS   American Association for Laboratory Animal Science
ACLAM   American College of Laboratory Animal Medicine
ALF   Animal Liberation Front
AWAR   Animal Welfare Act Regulations
CAAT   Center for Alternatives to Animal Testing
CDC   Centers for Disease Control
CIOMS   Council for International Organizations of Medical Sciences
DoD   Department of Defense
EPA   Environmental Protection Agency
FDA   Food and Drug Administration
GLP   Good Laboratory Practice regulations
HSUS   Humane Society of the United States
ICLAS   International Council for Laboratory Animal Science
IACUC   Institutional Animal Care and Use Committee
ILAR   Institute for Laboratory Animal Research
NC3Rs   National Centre for Replacement, Refinement, and Reduction of Animals in Research
NIH   National Institute of Health
OLAW   Office of Laboratory Animal Welfare
OSEH   Department of Occupational Safety & Environmental Health
PETA   People for the Ethical Treatment of Animals
PCRM   Physicians Committee for Responsible Medicine
PHS   Public Health Service
USDA   United States Department of Agriculture

Laboratory Acronyms

ABSL   Animal Biosafety Level
AV   Attending Veterinarian
Guide   Guide for the Care and Use of Laboratory Animals
HEPA   High Efficiency Particulate Air
IO   Institutional Official
IVC   Individual Ventilated Cage
LAA   Laboratory Animal Allergy
LAI   Laboratory Animal Infection
PI   Principal Investigator
PPE   Personal Protective Equipment
SDS   Safety Data Sheet
SOP   Standard Operating Procedures
SPF   Specific Pathogen Free

Glossary of Terms

Axenic   Animals raised under sterile conditions, uncontaminated by or associated with any other living organisms. Synonymous with germ free.
Bruce effect   A phenomenon in which pregnant mice abort or resorb their litter when exposed to the urine of an unknown male mouse.

Coprophagy Consumption of feces.

Dalila effect A condition common to mice, barbering or plucking of fur or whiskers from cagemates (hetero-barbering) or oneself (self-barbering).

Gavage Supplying a substance directly into the stomach via a small plastic feeding tube passed into the mouth.

Hypothesis A proposal intended to explain certain facts or observations.

Immunocompetent The ability of the body to produce a normal immune response when exposed to a pathogen or other antigen.

Immunocompromised The inability of the body to mount a normal immune response when encountering a pathogen or other antigen.

Reduction The use of methods which minimize animal use and enable researchers to obtain comparable results utilizing fewer animals or to obtain more information from the same number of animals, thereby reducing future use of animals.

Refinement Employing methods that improve the husbandry and performing procedures which minimize actual or potential pain, suffering, distress, or lasting harm and/or improve animal welfare in situations where the use of animals is unavoidable.

Replacement Employing methods which avoid or replace the use of animals in an area where animals would otherwise have been used.

Risk assessment A systematic process of evaluating the potential risks that may be involved in a projected activity or undertaking.

Sentience The capacity to feel, perceive, or experience pain.

Thermoneutral zone An environment that keeps body temperature at an optimum point to maintain homeostasis.

Thesis A dissertation on a particular subject in which one has done original research, often presented by a candidate for a diploma or degree.

Thigmotactic Movement of an organism that allows it to remain in contact with a solid body such as a wall.

Whitten effect A phenomenon in which the introduction of male mouse urine to female mice stimulate synchronous estrus in a female population.

Zoonosis A disease communicable between animals and humans.

Index